'Any book from Daniel Taylor is an occasion for celebration. No one knows or has thought more about the complex fusion of culture, landscape and spirit that is the very essence of the Himalaya. The story of the Yeti distills all of this magic into a single beam of illumination. And into this light walks Daniel, always the scientist, who reveals the truth, but in a way that only fires further the imaginations of all of us.'

—**Wade Davis**, Professor of Anthropology and Ecosystems at Risk, University of British Columbia, Canada

'Three decades have passed since I met Daniel Taylor ... Together we established a nature preserve as an abode for all creatures, mythical as the Yeti or real as bears ... Still people are calling 'Abominable Snowman' to the innocent Yeti who resides in the hearts of mountain people, who adds life in their arts and culture, and overall who helped them to raise their livelihood through tourism. I am thrilled Daniel has come up once again with ... the secrets of nature in the highest terrestrial ecosystem of this living planet.'

—**Tirtha B. Shrestha**, Life Member, Nepal Academy, Kathmandu, Nepal

'A mystery wrapped in an enigma wrapped in one hell of a yarn— and with big implications not just for protecting the Himalayas, but for the way we think about the world.'

—**Bill McKibben**, environmentalist, author of *Deep Economy: The Wealth of Communities and the Durable Future*, and founder of 350.org

'In one fell stroke of clarity, Daniel Taylor has re-written the history books and rid us of a decades old fantasy—that of an as-yet-undiscovered 'higher' being—higher at least in terms of altitude, if not the smarts to elude us for over a half century! The Yeti of the Himalaya (remember the cover of *Tintin in Tibet?*), Bigfoot or Sasquach of North America and the Mande Burung of Meghalaya—all have been put to rest in one go! Yeti slayer or eco-chronicler cum geographer extraordinaire—call him what you will, Daniel Taylor is a true Himalayan soul who has traversed the mountains from Arunachal to Uttarakhand—from a childhood in Landour to his more recent years spent in education and conservation, spanning a sixty-year swathe of time. He can speak to the people of the Himalaya and of the animals, with equal poise. He speaks as much to the city-dweller as to 'all who also make the wild part of their lives'. Indeed, over the years Daniel has made the high altitude wildernesses a part of his life! He takes us along on a journey from the wilds of India that he grew up in to the equally challenging wilds of the new age, in a particularly well-written series of lucid essays that are Thoreau-like at one level and Gerard Manley Hopkins-esque in their poetic thoughtfulness, at another. Unstoppable and unputdownable—you run the real risk of throwing it all away and heading to the mountains up north when you're done with the stories of Daniel's Himalayan lifetime as told in this book! Go for it ...'

—**Rupin Dang**, filmmaker, Wilderness Films India

YETI

One of the author's yak caravans, crossing Shao La Pass, the place where the Yeti was sighted in 1921 by the Everest Reconnaissance Expedition

Source: Author

YETI
The Ecology of a Mystery

Daniel C. Taylor

OXFORD
UNIVERSITY PRESS

Oxford University Press is a department of the University of Oxford.
It furthers the University's objective of excellence in research, scholarship,
and education by publishing worldwide. Oxford is a registered trademark of
Oxford University Press in the UK and in certain other countries.

Published in India by
Oxford University Press
2/11 Ground Floor, Ansari Road, Daryaganj, New Delhi 110 002, India

© Daniel C. Taylor

ISBN-13: 978-0-19-946938-3
ISBN-10: 0-19-946938-5

Typeset in Goudy Oldstyle Std 11/14.3
by Tranistics Data Technologies, Kolkata 700091
Printed in India by Replika Press Pvt. Ltd

Go and look behind the Ranges –
Something lost behind the Ranges.
Lost and waiting for you. Go!

So I went, worn out of patience;
never told my nearest neighbours –
Stole away with pack and ponies –
left 'em drinking in the town;
And the faith that moveth mountains
didn't seem to help my labours
As I faced the sheer main-ranges,
whipping up and leading down.

Till a voice, as bad as Conscience,
rang interminable changes
In one everlasting Whisper
day and night repeated – so:
'Something hidden. Go and find it. ...'

Rudyard Kipling

To the people of Shyakshila Village, Nepal, who protect the
Barun Valley ... and the Yeti

And to you who nurture the wild in your lives

Contents

Mysteries and Challenges

An Introduction

This story traces an arc from my boyhood in the Himalaya, chasing monkeys from my toys, through launching two national parks surrounding Mount Everest. What connects monkeys chased from toys and national parks around Mount Everest is the Yeti.

About this mystery of the Yeti, Himalayan legends assert that an animal much like humans inhabits the snows around Mount Everest. Turning the legends to serious inquiry are footprints. Stories do not make footprints—thus, if footprints exist, the maker of those footprints must also exist. And footprints that keep being found in the same basic structure across more than 100 years mean the evidence is not that of one aberrant animal's trail, for if the prints were made by a freak, after some years that individual would die and the distinctive tracks would stop being found. Moreover, the same basic footprint found in multiple sizes suggest a population that is reproducing.

This book clarifies what animal leaves these footprints. Various explanations were being advanced when I began the search. The prints might be made by an unknown wild man who for more than a century had eluded identification. They might be by an unknown animal but non-hominoid, related perhaps to the giant panda or gorilla with whom the footprints shared intriguing resemblance. A third explanation was the maker was known, but not yet recognized as able to make human-looking prints. And, of course, a fourth possibility—Yeti is a materialized spirit: not an animal, but a supernatural being, Science asserts, cannot, must not, exist.

Giving credence to the debate are always the footprints. That footprints were being found was never in doubt. Some 'thing' was making footprints. Still today, the mysterious prints continue to be found. Photographs are repeatedly taken. Thus, the Yeti is real, for imagined animals do not make footprints. Footprints are made only by real animals. So, as with Mount Everest, the Yeti is 'there', and as with those who sought to climb Mount Everest, I began a search to explain the footprints.

My search included discovering prints. It included tranquilizing a specimen of the footprint-maker and replicating in plaster of Paris the earlier unexplained footprints, especially the famous Shipton print from 1951. Solving that took from 1956 to 1983. I searched a 1,500-mile span of the Himalaya, the south slopes (India, Nepal, Bhutan) and the north slopes in China's Tibet. In dozens of wonderful expeditions, I visited almost every valley system across fifty years through that swath of mountains; I learned a range of local languages.

But, to my surprise, solving the footprint mystery did not answer the 'Yeti question'. While my discovery went worldwide in the news media in the 1980s, I realized the Yeti is a symbol of an idea as well as a real animal. My quest evolved. It investigated a concept greater than a Himalayan mystery: the question of humanity's relationship with the wild. By 'wild' I mean life uncontrolled by humans, that which makes Nature alive. For, aside from being a footprint-making animal, the Yeti is also an icon of wild humans.

I grew up in the 1950s along the edge of India's jungles. That was a wild from where animals attacked, an exciting world, and under my grandfather and father's tutelage I learned to navigate in it. That wild, though, like the jungles, is almost gone, and the search I ended up finding is of a new wild still with us. Life's threats now are human-made. You and me are what is dangerous for we reshaped the earlier wild to create a new—microbes our medicines do not cure, economies that are unpredictable, societies where people offer their bodies to carry bombs, a changing climate that changes everything.

Having left a Nature that once nurtured us but we could not control, we now enter a new reality which we also cannot control—one that gives true existential threat. Once we feared a natural wild, now what we should be fearing is a human-made wild. As the wilds change, the footprints of human existence show a consistent trail of our lives through the greater life that we never could control. This journey is the core dimension of being alive. In this glass-screen world we have now made, whose images may be real or 'special effects', beyond that glass screen, living in greater existence, rises a growing wild we did not intend and we most certainly have made.

For after explaining the footprints, and discovering vestiges of the old wild where the animal lives, my journey undertaken with many colleagues created national parks. We approached this not in the then customary mode of protecting species (in this case the Yeti and its jungle neighbours) but by finding a new approach with which national parks are managed. For if the nature of Nature has changed because of people, to live with this new, a new method is needed to ground human actions.

In Nepal and China's Tibet, lands between Earth's greatest human populations of India and China, was implemented a new way of living with Nature. Makalu–Barun National Park in Nepal, where the Yeti's secret was unmasked, pioneered this approach. That informed a year later Nepal's Annapurna Conservation Area; these two projects began a new preservation strategy for all of Nepal, led by Nepali scientists and officials, going beyond the approach where the army ran the parks. In the Tibet Autonomous Region of China adjacent to the Yeti's jungles, this approach trusted people even more. Qomolangma (Mt Everest) National Nature Preserve (QNNP) was established, then the largest protected area in Asia. The momentum went on with eighteen other parks across Tibet, some of which now are among the largest in the world. Together they can be said to protect 'the highest place on Earth for the highest need for the Earth'.

A human-centred approach to conservation was then near heretical (even though articulated a few years before in speeches at the Third World Conference on Protected Areas in Bali). Major conservation organizations and distinguished scientists had talked about the idea but had not yet implemented it. The need for a new participatory relationship with the wild is not just for park creation but global. With a people-centred approach, protection can engage conservation at larger, true planetary scale, and at the same time lower cost. As evidence has grown across a quarter century since then, the human-centred approach has become the accepted way.

By 1985, we had plaster casts to match the unexplained footprints. By 1992, the trail had established the national parks and protected the heart of the Himalaya. That personal story I described in an earlier book, *Something Hidden behind the Ranges* (San Francisco CA: Mercury House, 1995). I have written the story of national park creation, especially its extension across the Tibet Autonomous Region, in two other books: Chapter 20 of *Just and Lasting Change: When Communities Own Their Futures*, 2nd Edition (Baltimore MD: Johns Hopkins University Press, 2016). Also, Chapter 8 of *Empowerment on an Unstable Planet: From Seeds of Human Energy to a Scale of Global Change* (New York: Oxford University Press, 2012).

Still, though, the Yeti kept drawing me on, not as animal but as icon and idol. Icon and idol are powerful because of the 'I'—the I am who I am, the call of larger existence. This trail of

Yeti was exploring issues out of the three-million-year cavalcade of humans who everyday separate from the old wild. So, in 2010 I went again on a continuing Yeti search; it was fifty-five years after my first. I walked in now-familiar forever-protected valleys we had mapped during the park creation decades. In these personal quests in now protected valleys, I found wild animals to be more numerous and wilderness more vibrant than when we 'discovered' the valleys.

Is the Yeti a masquerade—yes, of some 'thing' needing explanation. Is the Yeti a mascot—yes, of a symbol of our evolutionary missing link. So, is the Yeti still a mystery—again yes, of awe. For in these pilgrimages I was touching where the planet rises uniquely into the heavens, walking within the greatest wildness on earth. Discovering understanding there happened because I 'went and looked behind the ranges, went worn out of patience, stole away with pack and ponies ... and the faith that moveth mountains.'

And today that call to understand can keep coming if we individually keep searching. Nature is beyond our control; life processes embrace forces bigger than DNA. Profound double helixes coil this invitation, interlocked spirals of myth coupled with science, leading to a definition of meaning that describes not protoplasm but the ground of all being: God, Brahma, Qi, Allah, Gaia, The Way, whatever. These forces give answer to those origins from where you and I have come, clarity to the lost and waiting ... something hidden, towards where we all must go.

When in nice coffee shops or city apartments temptation comes to believe what has been crafted for our living is a better place than Nature, this human grown world of cities and inside them our homes. But the truth is that manufactured environment still lives in Nature. Forces live still outside our shut doors, potencies so vibrant that a billion years ago they turned stark atoms into sentient life. Those forces are not gone that created the very nature of Life itself. They started the wild from some great energy that we can yet drink from. I might believe I am alive inside those doors, but do I truly live when I expect my days to be walked without getting feet muddy, or when I control microclimates up (or down) no matter how overheated the planet. No, the wild has never left; it is coming, indeed even through closed doors.

So, on those two expeditions when I went back into the Yeti's homeland on pilgrimage, I opened the door and stepped out, going into valleys protected. In going back into the old wild, searching the jungles and glaciers again, I walked beside the highest of mountains and deepest of valleys, searching for the wild man still living. I discovered something I had lost when as a child I played amid the wild. This book tells of that journey.

Conversations recorded here as conversations are drawn from memory. When those discussions occurred across the years of my quest, their lessons imprinted. In many instances, literal wording is retained in the quotes, for I keep a private diary in such times. In other instances, the words embody reflections from my field notes that encapsulate an event at that time but amplify meaning across time for how a story of a mystery I saw in the newspaper at age eleven led to larger learning about the challenges of life.

As the reader enters this story, I make two introductions. First, let me introduce the animal 'Yeti', but now spoken of as a named animal. Yeti in this book is a persona with triple identity: animal, mystery, and icon. Hence, it is written with a capitalized first letter. Additionally, I introduce the land, the Himalaya, where Yeti leaves its footprints, a place, *the abode of snow* (Sanskrit: *hima* = *frost* and *laya* = *dwelling*). When the British came, they fixated on the mountains, turning the name to Himalayas, summits for further conquests. But to the people who had long lived in the heat at their base this was a place, a renewing dwelling where pilgrims went, where animated spirits lived, and life cures could be found (such as those by the Yeti-like Hanuman). This book, in such understanding, speaks of the Himalaya, the place: not the Himalayas, the mountains.

Into this mystery and magic the reader is invited. In this quest is some 'thing' (the Yeti) that is more than the 'I' (you, the reader, as well as me, the seeker). That animation and animal calls: 'In one everlasting Whisper, day and night repeated: Something hidden. Go and find it.'

one

Arriving at the Yeti's Jungle

February 1983. I stand at the entrance of the Barun Valley. 'No jungle in Nepal is as wild as the Barun.' His Majesty, the king of Nepal, had told me two years earlier. His advice was no surprise, for it was here, ten years before, that Ted Cronin and Jeff McNeely had made their Yeti discovery on the ridge above. Since their discovery, no other Yeti hunters have explored these jungles. Permissions were not denied to them, but none others asked. Searches had concentrated in the high snows where Eric Shipton had taken his iconic photograph of the abominable snowman footprint in 1951.

His Majesty had told me, 'The Yeti may be called a snowman, but whatever makes the snow footprints, that animal must live where it can hide. And of all my jungles the Barun is the wildest. If I were Yeti hunting, that is where I would go.' The king was smoking his pipe as we sat in his study at the end of a palace workday, our feet resting on the pelt of a snow leopard. We had become friends in graduate school. The king went on, 'And, if for some reason the Yeti needs snow, few glaciers are as isolated as those that feed the Barun River. Snow there constantly blows off Everest, Lhotse, and Makalu creating huge glaciers.'

As I look down at the Barun River, marks on the gorge below show little fluctuation in the water levels. This is a surprise. Usually, river volumes in the Himalaya surge a hundred-fold with the monsoon. But the marks below do not show a great change of height. So, about to enter the valley, I am already finding mysteries.

'Full of bears and bamboo,' Ted had described the Barun when I'd asked on the phone. I was pressing for information. He went on, 'Unbelievably hard to make progress. No trails in that jungle. We hacked our way for days, and got only miles.'

In my talk with Ted, I had found it hard to believe that no trails existed, for trails are everywhere in the Himalaya. But as I look into the valley, I do not see where a trail might run. Cliffs block the way. To get around them the route might climb to the ridge and enter mid-valley. My altimeter says 3,100 feet here, and my map shows near ridge tops at 10,000 feet rising to 26,000 feet where they connect to Makalu, Lhotse, and then Everest.

Makalu, the closest of these summits and the world's fifth highest, is only 15 miles away. But all before me is an extremely dense jungle. A weather phenomenon occurs in the Barun and in the valley just to its north that happens nowhere else. This is the only place where the earth rises into the jet stream that races around the planet. As it whips off Everest, Lhotse, and Makalu, it creates a low-pressure zone. That pulls the warm air from India up the Arun Valley, one of the deepest valleys in the world. The air turns where I stand and goes up the Barun. When it hits the chill from Everest, Lhotse, and Makalu, it snows, snowing more than anywhere else in the Himalaya. The snow is not coming from the moisture in the sky but from the air sucked up from below.

Walking back to camp, I pay off our porters. It took our team five days to get here, but at the pace the porters depart, running down the trail, their families will greet them tomorrow. Around me now is my expedition, not intrepid 'Everesters', not renowned scientists, but my family: my wife Jennifer, her twenty-one-year-old brother Nick, and my two-year-old son Jesse. Two others, Bob and Linda Fleming, will join in ten days. Bob is the leading natural historian of the Himalaya. Together, he and I have been planning this expedition for three years. I've been on the Yeti's trail for twenty-seven years.

And there is another, now-evident puzzle about the valley. Here, at the confluence of the Barun and Arun rivers, up and down the Arun Valley, we see villages with their terraces high on these slopes, slopes that cannot squeeze in another field, and homes obviously distant from water. With so many people packed on to what appears to be very difficult Arun slopes, why is there not one field in the lush Barun? Indeed, no trail? The whole Arun Valley is domesticated. The lush Barun has no homes. What keeps even local people out?

The next morning, from our tent I hear Pasang kindling the cook fire. Stepping into the morning, on a rock 50 yards away I see a man dressed in green, a muzzle-loading rifle beside him. His eyes, careful and curious, are uncovering the unknowns of our camp. Holding out a mug to him, I motion him to come down. Pasang blows on the coals of the cook fire. With no motion in the bushes, a minute later the stranger strides into the camp. Where's the muzzle-loader?

My eyes can't leave his feet. Through years in the Himalaya I've seen feet thick with calluses, toes splayed like a bird's claws, and bare feet forming to ground the instant they touch. These feet, though, are different. Toes slipped aside pebbles and reached the solid stone underneath, showing a personality spontaneously seeking a secure footing.

We squat close to the fire. Nestled in the coals, Pasang's teapot gurgles, and in the pot tea leaves swirl. The newcomer watches as Pasang pours milk powder into a mug and then ostentatiously adds a double serving of sugar. We are three: a man who owns a muzzle-loader, therefore a poacher; a former Tibetan nomad now working as a cook; and a white hunter seeking some animal. The poacher's toes, though squatting, keep moving—is he poised for action or just nervous?

My brother-in-law Nick exits the tent. Pasang stirs up sweet milky tea for Nick, himself, and for me.

'Where is your village?' I ask the hunter.

'Shyakshila.'

'Where's that?'

'There, on the ridge.'

We are silent as the pot gurgles.

'What is your name, my jungle friend?'

'Lendoop.'

'What are you doing so early in the morning?'

'Oh, I've just been walking.'

'Why do you carry a gun, Lendoop?

'I don't.'

'What was that you had then a few minutes ago on that rock?'

'Why have you people come here?'

'We are looking for animals in the Barun Valley.'

'You won't find them.'

'Why?'

'Because you cannot go into the Barun.'

'But we have permits to go.'

'That doesn't matter, you still won't go.'

'Why?'

'There is no trail. White folks cannot walk without a trail.'

'Why?'

'Because there is no trail.'

'Can we make a trail? It is only a jungle.'

'You won't know where to put the trail.'

'Can we find people who will know, who can help us make it?'

'How much will you pay?'

'Twenty rupees per day.'

'You won't find anybody.'

'How much is required?'

'Thirty-five rupees per day.'

'Maybe we'll pay twenty-five.'

Lendoop looks into his mug. It is empty. Pasang fills it. Lendoop smiles with the second helping of sugar. Pasang adds a third.

Casually, I resume my questioning. 'What animals are there in the jungle?'

'All types.'

'Like what types?'

'*Serow*.'

'What else?'

'*Ghoral*.'

'Others?'

'Red panda, leopard, wild boar, barking deer, musk deer, bear, snow leopard, *thar*, three types of monkeys.'

'Blue sheep?'

'No, those animals are found in Thumdum, two valleys to the east.'

'Any jackals?'

'Sometimes, not always.'

'What animals do you hunt?'

'I don't hunt.'

'Are there any small cats in the jungle?'

'Yes, two types of jungle cats.'

'Any fish in the river?'

'In the Arun, yes. In the Barun, no. The Barun is too fast. It is a dangerous river.'

'Are there wild men in the jungle, *bun manchi*?'

'No.'

Lendoop looks at the bottom of his cup. Pasang refills it, and adds the two spoonfuls of sugar.

'Any villages in the Barun Valley?'

'How can there be villages? I told you there are no trails,' he says scornfully.

'Do you ever go into the jungle?'

'No, there is no need.'

'How do you get your meat to eat?'

'We kill a sheep or goat.'

'Isn't it cheaper to shoot a wild animal with your gun?'

But Lendoop knows what I want, and so finally, he breaks off a twig and a man whom the outside world would call illiterate, draws arching lines in the dirt. Pointing to the longest, he says, 'This is the Barun. Big mountains are here, high above the valley, all ice and snow. Makalu here, another big mountain, Lhotse, here.' Slanting lines run through the dirt. Check marks are going down some of them. 'Another river, the Mangrwa, joins here. At the top of the Mangrwa are no big mountains but a pass into Tibet and China, very easy to cross, where there are no border guards. After where the Mangrwa and Barun join are streams, like this, the Hinju and the Payrenee.'

The sketch is simple, also different from any US, British, or even secret Indian Army map, all of which I've studied. In the dirt is accurate cartography of the Barun, and it looks like it's done with consistency of scale.

'How long would it take to walk to where the Mangrwa and Barun meet?'

'You cannot. There is not a trail.'

'How long does it take you?'

'One day.'

'How long if you take us?'

'If you are strong and we make a trail, three days.'

'Three days? We can walk to Khandbari in three days.'

'No. It took you four days to get from Khandbari to here, and one day before that to get to Khandbari.'

He is right. Our movements must have been communicated through the valley, particularly among people walking the Arun given the questions I had been asking.

'Do people ever find the *Shockpa* or the *po gamo* in the jungle?'

'No,' says Lendoop. 'The Shockpa is not an animal.' The statement is like ice-cold water on a hot day. Suddenly, Lendoop stands, hands his teacup, and walks out through the bush. Is that terse statement the end?

All this way, and now we learn that the animal doesn't exist. For years, I've kept saying that the villagers would know once we came to the right place. I've been saying that we would have to spend days just listening, that we shouldn't rush into the jungle, that we should avoid searching ourselves when they've searched across generations. Nick heard the question just now. His Nepali is good enough to understand Lendoop's 'No'.

Grandpa always said, 'Listen to the local people; they know their jungles.' It was on Grandpa's dining-room table, twenty-seven years earlier, that I had first met the Yeti staring up at my eleven-year-old's curiosity. Eight years before that when I was three, he and my father had started taking me into the jungle. We tracked tigers to protect villagers, shot wild boar for meat, but, most importantly, the jungle became a familiar place where we moved in comfort and care.

'It's all part of the quest ahead,' Nick says, breaking my reverie. 'Some animal made Cronin and McNeely's footprints. The place where they camped that night is just above us on that ridge. The Yeti may or may not exist, but footprints do. Something walked across the ridge above.'

Maps were made long before the written word came into use. For 10,000 years maps have explained the human experience, a form of expression nearly as old as human painting, millennia older than the use of written words. Making and understanding maps is the first literacy used by humans.

Cave-wall maps accompany graphics of great hunts showing clearly that 'we went there and got this'. Millennial-old maps combine the earth with graphics of the stars—'I am here located by this out there'. To define place, from the time when our species came out of the wild people used this ability to connect with place.

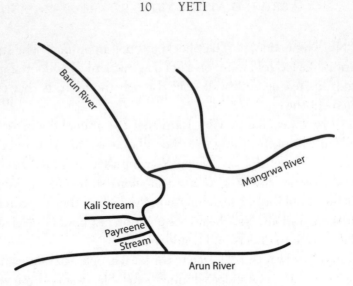

1.2   Lendoop's Sketch of the Barun Valley System

Source: Copied from Author's Field Notes

People then guided their travel as individuals today use GPS. Locating resources, they described the world. Early Greek maps from Anaximander in the seventh century before the Common Era show that the farther one ventures from the civilized world (Greece), the wilder the people encountered. At the fringes of Anaximander's map people are animal-like. Homer's two epics, the *Iliad* and the *Odyssey*, confirm this view.

Maps centre logically on where people consider home to be. When the British claimed their empire, presuming their definition of the world, they prescribed the centre for all maps through the foyer of their Greenwich observatory as the prime meridian. When Pei Xiu, the great Chinese cartographer in the third century of the Common Era, made his maps, he followed the same practice using grids that expanded across the Chinese empire. The great Arab maps in the ninth century placed the centring meridian, the *qibla*, as the direction towards a place (Mecca) rather than from a place.

Maps' purpose is to shrink the world into comprehension, and even 4,000 years ago they flattened three dimensions into two.

Before people could fly, maps gave people a bird's-eye view. For 2,000 years, they were doing this for Europe, China, India, Africa, and the Middle East. Before mathematicians, mapmakers solved the challenge of proportionality, shrinking the world according to scale (an inch, for example, to indicate a mile). Representing the world in two dimensions allowed it to be rolled up, taken under one's arm—this mastery of transferable knowledge (essentially a portable library) dates back 2,000 years, long before we had computers.

Layers of information can be conveyed: directions for travel (here is the pass, the crossing of a river), location of resources (timber or minerals or fields), dangers (an enemy to be avoided, a swamp). When headed into the unknown, and the world a confusion of possibly hostile people, plants, mountains, and rivers, a map organizes the uncertainty.

Mapmaking skills have grown since these early beginnings. Two-and-a-half millennia ago, Aristotle proved the earth was round, a key insight. A ship could be seen rising over the horizon, first the mast tip, then the whole mast, and finally the ship. Visibly evident is the ship climbing over the curvature of the earth. And from this came the understanding about how to shape the world into two dimensions. Watch a lunar eclipse. The sun is obviously round, and the moon also, then in eclipse can be seen the rounded shadow of the earth crossing the moon. And finally, Aristotle noted, some stars can be seen only from certain parts of the earth—if the earth were flat, the heavens would look the same from everywhere. And so the world's shape came to be understood; the round earth became flat ... yet it was understood to be round.

Working to make more accurate maps, Eratosthenes, two centuries later, calculated the circumference of the round earth (to half-a-per cent accuracy). He took the lengths of shadows off known heights in different parts of Egypt. If a 100-foot tower at one place gave an 80-foot shadow at one location, and, at exactly the same time, that same height gave an 81-foot shadow at another location, the reason was the earth's curvature. Eratosthenes went on to postulate the sun's consequences on a round earth's habitats. On a round sphere the central band would be warm where it received maximal sun, on either side temperate, while the temperatures were increasingly frigid at the poles—he originated the term geography.

But maps of the Himalaya have been a problem. Two-hundred years ago the British started mapping the Indian subcontinent. Beginning at the Indian Ocean they built towers, sighting and measuring tower-to-tower, demarcating lands, deserts, jungles, even the tops of mountains they could not climb. Though they could measure the Himalaya, they were denied entry. So the Great Trigonometric Survey sent in spies. In prayer wheels were compasses, and the papers in these prayer wheels were draft maps, not prayers. The prayer-bead strings had a hundred beads (not the usual 108 for each blessed name of the Buddha). With each stride, the spy flicked a bead, and when a round of hundred passed, one of the ten beads hanging off the end was flicked. Valley after valley was mapped. But no spy ever attempted the Barun.

Before coming I had checked the Barun Valley on every map I could find. Except for the king, I could find no one who had flown over this valley and studied what he saw. So the Barun maps that I studied had been built from guesses. The usual guess for the elevation of the river confluence here with the Arun was 3,700 feet (so now before entering the valley, I know the maps are off by a factor of one in seven). Guessing also that at Makalu's base, it was determined that the river falls 11,000 feet in 15 miles. A tributary was postulated to join from the north. Lendoop has now told me it is called the Mangrwa.

But everything was unknown for what must be 200 square kilometres of jungle inside the valley. Does the river bend? Probably, but a river losing 11,000 feet in 15 miles will move very fast. So maybe there are no bends. Standing at its entrance, we do not know what waits inside the valley—except dense jungle and mountain slopes rising to the highest places on earth.

☙

THOUGH WE STATE THE METHOD DIFFERENTLY, shared respect for village perceptions is what a self-appointed, scientifically marginalized group terms the cryptozoological method, the science of learning that about animals which traditional peoples already know. Richard Greenwell, the secretary of the International Society of Cryptozoology, advised me before we departed, 'Dig for information hidden in local knowledge. Sometimes this knowledge coincides with our own ... sometimes it doesn't. The fun of cryptozoology is trying to determine if the native peoples are right or wrong in ways science will accept.'

That evening two hunters came into the camp and accepted our offer of tea. They always hunt together, they said, because they share a muzzle-loader. Eleven years before, they were stalking ghorals along a stream bed, hoping to get close to the animals while the noise of the stream covered their sounds. By the stream's edge these men saw two baby shockpas splashing water on each other. Their fur was grey, and it lay both ways on their bodies.

'Shockpa' is the local name we deduce for Yeti. Other informers come, and over their cups of tea, reports of shockpa's long vibrating screams follow. Other visitors deny po gamo or shockpa. Others claim shockpas are ghosts. Some others bring no jungle tales but come out of curiosity and, of course, for the tea. Nearly all seem afraid of the jungle, entering in as little as possible even though they live on its edge.

Only the hunters who saw the two shockpa babies possess Lendoop's aura of jungle savvy. They return on the fourth evening and want money for their story. The details fit with the *po gamo–needeene* themes I had heard for years. The two hunters ask me to climb the hill on the other side of the Barun River, their village's side, not Shyakshila's, and from that ridge they point to a peak above the narrow gorge that is the mouth of the Barun Valley. 'The name of that peak,' both men assert, 'is the Shockpa Summit; many shockpas live there. The river is on the other side.' But when I ask these two men whether they'll take us to the river where they saw the babies playing, they say no. They never return.

Nick points out that among the accounts we've heard, no two shockpa descriptions match, whereas descriptions of other wildlife do. Visitors to the camp from both sides of the Barun (Sibrung and Shyakshila villages) agree about the Shockpa Summit as the peak guarding the valley.

Each morning, though, Lendoop returns; his descriptions are consistent. His observational powers are proven in how he notices all that is happening in the camp. And as he starts understanding what is happening, voluntarily, he starts pitching in to help. Accompanying him after the first day is Myang, one of Shyakshila Village's headmen. After four days, no other person is willing to take us into the jungle. Lendoop and Myang have, it appears, staked out their territory as our guides. How are such decisions passed through village hierarchies? Presumably, Sibrung and Shimong, two neighbouring villages, also have hunters—but their hunters have never even come to our fire.

Our tents are pitched in the middle of a trade route that has linked India with China for a thousand or more years. The spot is known as Barun Bazaar, though it is just flat bare land; the only building being a stone shelter, but all tell us that here regional markets occur periodically and twice a year there is a festival to celebrate the pure Barun. People who travel along this route are on practical missions up and down the Arun Valley, leaving homes to make small profits at some market or the other. Life doesn't offer them the cushion of time we're experiencing of four days just sitting, gathering stories.

During the days our camp sits astride this trail where every trader, pilgrim, or messenger is a potential informant. We try not to show particular interest in Yeti stories. As the reports mount of what they describe about the dense jungle valley behind the camp, what is confirmed again and again is that only a few know that jungle. What the king says is the most wild jungle in Nepal seems to be a place everyone stays away from, even those who live on its borders. They shudder when we suggest that they might want to make a home and fields inside. Only to the jungle's edge do they go, seldom past the Shockpa Summit, to gather medicinal herbs, grass for animals, bamboo, and timber.

Many who come are curious about us, especially women and children. These people are not the wayfarers who stop for a moment. These are mostly women, and as women do throughout the Indian subcontinent, they come in groups, and they come to see one thing. On the fifth morning, we're awakened by quiet whispers on the other side of the thin tent walls. 'Jesse.' 'Jesse.' 'Where is he?' 'Is Jesse awake?' The women wait.

Jesse steps out of the tent, pushed by Nick. From inside we hear, 'This is Jesse.' Peeping, we watch Jesse walk towards them, then run after something he's spotted, with a bobbing stride that women universally seem to enjoy. He is embraced by their gaze.

'His clothes, have you ever seen such brown, green, and yellow?'

'Look at how big he is! I am told he is two years.'

'The yellow hair, the sun shines in it like straw after harvest.'

After breakfast Jesse has his bath as he has each day of this expedition. With so many sights and events to explore, this two-year-old gets dirty quickly. Over the fire Pasang heats water in his giant pot. Sitting in a small yellow plastic basin, Jesse splashes. Jennifer soaps him down with the first tubful, then rinses him with a second. Such copious use of warm water and soap is new to these women. More than hair or clothes, Jesse's cleanliness sets him apart.

Today, though, at last, we are ready to start the trek into the jungle. As tents go down, a trader we have never seen before walks into the camp. He looks around, sees Jesse in his bath, and says, 'That must be Jesse.' Like the Yeti for us, Jesse gains legendary status. It is indeed time to head into the jungle.

two

In the Yeti's Jungle

Previous page:
2.1 The Yeti's Footprint in Dumjanje Stone

Source: Author

I hoist Jesse on to my shoulders, and we start the climb into the Barun gorge. He steers me by the pressure of his knees on my chin; for him, it is easier than pulling my ears as before. My head falls forward to loosen his pressure on my windpipe, making me understand how horses get broken; it's also how parents get made.

Contrasted to his swaying seat on the trek in, now he has lithe balance. Firmness of seat and his knees under my chin hold him, so I no longer must grasp his legs. Each of us brings skills out of childhood. Balance is one such skill which is often overlooked. Jesse and I had practised in America with child-carrying packs, those with 'safe' straps and padding, that made the weight easier on the father. But positioning his head right behind mine caused him to lose the view of where we were going, missing the chance to think about guiding our way, or to spot a squirrel in a tree. A two-year-old sees excitement in a squirrel's jump.

So now I am seeing the world again through eyes that are two-years-old. Such seeing had opened India's jungles to me when England still 'ruled' India. I was then bursting with energy, and to spend that the jungle was just outside our door, a world into which I ran daily. It taught me to learn patterns different from the linearity of A, B, C, or of symmetrical wooden blocks. Running into the jungle initiated me into the world whose ways work by complexity; it taught me to wrestle with questions. Unlike school, the world presents us with few questions that have right answers.

Jesse will now start learning how to enter the jungle, learning to wrestle with questions rather than believe that the world leads linearly to answers. Learning through balancing. As two eyes give depth perception, four eyes (of young and of old) give perspective. The human journey, with the changes of the new human-made age now upon us, goes into a new wild where life functions not by rules, but through balancing of relationships. Safety is wanted both today and also in growing skills for tomorrow. When 'the wild' becomes familiar in this new age, what is found is 'the alive' of the new today. As with learning languages at early ages ... shaped also at an early age is how the mind makes connections, setting values in place of how to engage Nature that remain with us as we grow.

We catch up with Jennifer an hour out. Jesse's knees come forward into my chin, his signal to stop. 'Mama, on the ground, I need that stick. I need it.' Dutifully, she hands up the stick. A memory flashes of the stick I took when I went out; it stood by the door of our bungalow in the Himalaya. I felt magical powers when I held it, a stick that did a double duty swung up to my shoulder to mimic a gun. Each time I came back, I kept it by the door, near Grandpa's staff. Swinging his stick now, Jesse directs the route up.

With the improbable purpose of heading into the jungle to seek the Yeti, the question hanging is whether to take our two-year-old into the jungle? To partly pass my childhood to my son is of value. But is it safe? DNA may carry our bodies, and values carry in what we share. All that is a reason, but is it safe? My father grew up in the jungles where the Ganges comes out of the Himalaya. Home, until he left for college and medical school in America, was in tents that moved each week by ox cart to a new camp, and under the flaps of the big tents came in jackals, snakes, rabbits, even a she-wolf that tried to lift him from his crib. Dad learned to understand the calls of birds and see signposts in trees, and developed an ability to find his way home following his instinct rather than trails. When I was two, Dad and Mother brought me to the Indian Himalaya. I want Jesse to imprint Nature's cadences and tones too, with mind-shaping self-definition as with a mother tongue, understanding being with the wild and being comfortable there.

But is it safe? This question will be answered only at the end, as with the quest we now take for the Yeti. The great reward of exploration is not knowing beforehand what will happen.

Stepping off the trail and walking out on to a crag, Jesse slides off my shoulders. Across the gorge he spots langurs. 'Most monkeys do not eat leaves,' I explain. 'See the long tails on these, this helps them balance as they leap in the trees seeking the right leaves.' Jennifer and Nick catch up, accompanied by Lendoop, muzzle-loader on shoulder. Strung out behind are six porters carrying fifty-pound loads.

Lendoop leads us now as we leave Shyakshila's fields and enter the jungle. I lengthen my stride to catch up, but he then speeds up, calling back: 'The trail is good. This way'. Each day I grow more attracted to this man. That morning when he showed up with two unexpected porters, I had ribbed him saying, 'Adding two more jobs for your neighbours', as he insisted on lighter loads. His reply had been, 'They are my friends, it is your job to help.'

We've now climbed above the cliff that blocks the Barun Valley. It seems there is a way to enter the valley along its crest without having to climb above the next set of cliffs. Ten-foot tall grass surrounds us. What happened to the trees that must have once been here? No sign suggests an attempt to make fields. Turning around a bulge from where the slope levels, sits Lendoop in his green hunting clothes, arms akimbo on knees, like a frog on his rock. Grinning, he croaks: 'This place is the camp. We stop now.'

'Impossible. The day is only half gone. You and the porters are getting five more rupees per day than the government rate and are carrying light loads. We must go inside the jungle.'

'Porters could go, but baby cannot. We must camp here. This is the last place where tents can be set up. After this, there is no flat place for one long day.'

As evening falls, Jennifer, Nick, and I sit outside our tent, the stillness settling. Nearby, Jesse helps Pasang push sticks into the cook fire, knowing now not to pull them out once lit, watching as they burn, and then pushing the non-burning end. A croaking sound comes from the porters' camp. Nick returns after investigating it, his face a mix of grin and grimace: 'They caught frogs in the stream we passed back there and are now doing a live roast.'

As night wraps around us, tents sparkle as dew settles in drops on the fabric with a reflected sunset. The wind plays in the bows of the trees. With the darkening day, birds wish each other goodnight, interspersed with an occasional mysterious hoop. As animals go to rest, the day celebrates the night's coming. The Barun River below echoes a steady roar.

༺༻

AS MORNING LIGHTS THE OPPOSITE SLOPE, I tighten a high-powered scope to the tripod's swivel. Contour after contour on the south-facing slope is scanned, seeking animals coming into warmth after a winter night. The ghoral should be coming out, maybe also the serow. But yesterday and today, only langurs were seen. Twenty-seven years ago, when I had my first sighting of Yeti footprints in a photograph in the *Statesman* newspaper, the accompanying text said that a curator at the British Museum suggested that a langur was the maker of the mysterious footprint. Now in the Barun, langurs are the only wildlife seen.

Since childhood, I have been trained to start the hunt before dawn and stay quiet in the camp. But since dawn, porters have been noisily coming for cups of tea to our campfire with no sign of preparing to depart. But I hold my tongue; a working relationship with them and Lendoop is more valuable than any result for today. Finally, the hunter hands Pasang his empty mug and walks to the jungle edge. The others hasten to their loads.

With this orchestration of the morning, Lendoop has let all know who leads this trip. Without that leadership coming from all his team's desire to participate we will not find our quarry in the jungle. He is now swinging his kukri through the bamboo. The barrier falls with that large curved Nepali knife, stalk on to stalks before, each stalk an inch thick, stacking each stalk by the same kukri swing that cuts. How do bears get through this? For it is out of this bamboo, Lendoop has said, that they come to raid Shyakshila's fields.

An hour later we cross the ridge we've been climbing and drop to the west-facing side. Lendoop leads among gigantic trees with a maze of old dead logs and branches where decaying vegetation lies underneath like a sponge. I thrust my walking staff; nearly a foot deep. Looking ahead, I suggest the way seems easier above. 'This is the way,' Lendoop replies. 'There might be open space along the base of that cliff,' I say. 'This is the way' is all he says.

At times we cross over trickles that sound beneath the decay. The slopes are usually dry in the Himalaya at this time of the year. But as we enter the wet air of the jungles, there's periodic gurgling under the jungle floor. Jesse hears it too. 'Water, Papa. I want to drink some water.' Why is water running long after the monsoon? A steep slope like this should drain fast. Again, I thrust my walking stick in. Again, the forest humus is a foot deep—never have I seen forest decay so deep. It is, of course, this decay that is soaking up rainwater and then gradually letting it out.

I keep looking for footprints. Lendoop claims he's seeing tracks, calling out a serow once, another time a bear, but the bent blades or broken twigs he must be connecting into shapes are beyond my detection. In jungles such as this it grows more evident how hard it will be to see animals, made harder as all animals except monkeys are shy. Grandpa taught me to use streambeds for a wildlife census. But in the absence of visible streambeds that technique is impossible. I will need a winter storm to bring snow low to spot footprints.

But then at the place Lendoop calls Dumjanje, a somewhat level area, he says, 'See this. You have been asking about Yeti and Shockpa.' It is the first time I've heard him use the name 'Yeti', and when I used it back at the confluence he seemed not to know what I meant. 'See this. Here I am showing you a Yeti footprint.'

I looked at the large flat rock to which he was pointing. Unmistakable, embedded into the stone is a large footprint with bulbous toes, exceedingly similar to the footprint that Shipton photographed in 1951. Beside that print is another, looking very much like a person's hand.

'You asked me about Yeti footprints five days ago—I am showing you,' Lendoop says with a wide smile. 'We camp here.' It is clear that Lendoop knows this footprint is not animal-made. Then he goes on to say, 'Signs are on the land that people do not understand. Forces that make footprints and handprints into stone must be very strong.'

ଉ୭

THE NEXT DAY, WE CROSS PAYREENI KHOLA, and the cliff scraped bare leading to the stream shows that even tree and bushes are moved when the cliff above breaks.

A roar has been growing. First, the sounds were whispering like the wind. Now, the air is cold, plants drip with spray, and the sound grows. Jesse's knees pinch my ears. Pushing apart wet bamboo I face a shaft of rock. Hurtling down the shaft a waterfall has pounded out a pool of the rock. Water swirls peacefully in this pool, then slops over the outer ledge and splashes down, another waterfall with more spray.

The Kali Khola stream has carved this raceway into a stone cliff. I look up the stream's opened slope framed with green jungle. A rainbow glistens against the blue sky. Since our first meeting, Lendoop has spoken of this pool where the water slops over the narrow ledge, the only place on the Barun's slope to cross the Kali Khola. That stone lip is 20 yards beyond my boots. To arrive at this spot I had not sighted a single trail mark. Shapes guiding Lendoop were of Nature, not man, such as tree blazes; he was reading a language in the deep intricacies.

Lendoop stands in the pool. To join, I must step from where I stand, then cross along the rock edge. I take the first step, left hand stretched so that my fingers grip what seems to be a bush growing out of the wall, placing the right hand with the walking pole on the rock. Jesse's arms embrace my forehead, knees tighten. Surprised to see us step out, Lendoop shouts, 'Sahib, sahib, don't!' and charges towards us.

Jesse flings back from Lendoop's charge, still holding my head, pulling me off balance. My left foot slips on the ledge as my right hand lets go of the walking pole to grab his leg. The left foot sliding and the right holding, my left hand holds the bush that grows from the rock. I have spun halfway around, and my turned head sees Jennifer stepping out of the bamboo.

Jesse's mother's face is a universe of expression—mouth open, a silent scream as she watches her son, who, feeling my fall, has held firm and locked his knees even more. Throughout, the Kali Khola never misses a splash. Bush roots deep in the well-watered rock hold me, and my left foot is secure again.

'Sahib, dangerous, very dangerous. Me, I carry Jesse.' Lendoop stands beside me. Adrenaline races. I storm past, across the stone ledge. With the magic of a mountain guide, he melts into the rock so as not to take up space as I pass. When in the pool between the waterfalls, I look back at Jesse's mother, her eyes still wide.

ᘜᕲ

2.2   The Cascading Waterfalls of the Kali Khola

Source: Author

A WORD ABOUT SAFETY IN THE WILD. Rock climbing is safe because of the rope. Footholds may slip, handholds let go, but with a belay comes safety. Danger in the wild is almost always from falling. The risk of infections requires proximity to contagion, and where few people live is little toxin. Animals may attack—but that happens mostly in the imagination. In the wild, the major danger is a slip. As a belay arrests a fall, a tool to prevent falls is a walking stick. With the stick, the normally two-footed human animal steps back evolutionary to three feet. And, as just happened on the wet ledge, it is possible to have a fourth leg also. Putting my hand into the bush to use all my four feet was a reflex learnt early in my jungle childhood. Trusting my partner on my shoulders was another skill learnt. But there are always surprises, such as Lendoop's mistrust of the non-jungle man.

Did early hominoids use walking sticks as they first began to hobble on two feet from four? (The first steps of *Australopithecus* must have been teetering, probably needing some help.) As a young boy I carried my special stick. If a wild animal attacks, having the staff can also serve as club or, when sharp-pointed, can serve as a lance. What tools do Yetis use? I've often wondered.

☙

HOURS LATER, CLIMBING A STEEP SLOPE, we enter a clearing, the closest approximation to flat ground in three days, a hundred times bigger than the depression between the roots of the giant oak where we unfolded the tent on our second night. Here between splendid chestnuts, at 7,250 feet, is our mountain camp. Jennifer christens it 'Makalu Jungli Hot'l'.

The sun shines through the canopy. Ticks are out too, moving very fast. As we plan camp, one homesteads in my crotch. On Nick they find soft tissue. Tents go up, where Jesse retreats inside. All of us pull a double layer of socks over pant cuffs. The insect repellent is sprayed up and down pant legs *and* the tent zipper. Jennifer and Jesse sit inside reading *The Cat in the Hat Comes Back*.

As the temperature drops with the arrival of night, the ticks nestle underground in the warmth of the decaying humus. Around the cook fire we sit on stools, a jungle luxury carried in thanks to the porters. Pasang breaks the routine of rice and lentils and serves spaghetti. Turnips and beans get stewed too.

The spectacular *Petaurista Magnificus*, Hodgson's flying squirrel, calls from above, an animal of pristine forests, one-metre long, having a maroon coat with a yellow stripe. Its calls continue as it glides over the camp, seeking berries, insects, fungi, nuts, even dead animals for its dinner. Ever descending in repeated glides it explores the lower slopes. Each dawn it returns uphill by climbing a tree, gliding almost horizontally back to the hill, finding another tree, climbing that, gliding again; each time a tree-height higher than before on the slope, ascending this way while spending minimal time on the ground where predators might lurk. When the sun rises, the squirrel will be back in its hole in some tree above.

As we lie in our sleeping bags, out of the darkness comes the call of the tawny wood owl, the highest-altitude Himalayan forest owl. Appropriately mysterious, its *hoo-hoo* (the second *hoo* dropping in pitch) resonates through the dark. One-and-a-half thousand feet below, the Barun River vibrates the valley, dropping its approximately 1,000 feet of elevation every mile. The night air descends clear and cold from the glaciers of Makalu, Lhotse, and Everest as we lie in jungles circled by these three of the world's five highest mountains.

We are where we had planned to come: just below the snowline, at a moderate altitude, in the uncharted Barun jungle, and on a north-facing slope where we can, we hope, observe animals across the valley as they come out on the south-facing sunny slopes. That is our hope—to see the hyper-shy 'something' across the valley.

Our walk in showed plenty of food in the trees for the 'something' to eat—if we only knew what it ate. But the ticks signal that we did not come to a no-human world. Ticks are thick because herders stop with their goats en route to alpine meadows on the ridge above. Except for these humans, wild animals have possessed this valley.

As I listen to the night talk, I ask myself again: might there endure another wild man too? Forty million years ago when these mountains began rising from the sea, what animals were here— that would have been thirty-seven million years before hominoids were anywhere on our earth? Could there have been pre-hominoid in this place? It was, of course, a dense jungle then, but nearly flat, for forty million years ago was when the Himalaya were starting to ascend. People, so far as is known, first walked the Arun Valley 2,500 years ago, coming up this river that predates the Himalaya, and the waterway would have been a corridor along which animals moved. Perhaps a pocket of pre-hominoids, while humans advanced in the outside world, became secluded in this valley protected by the wilds we now sleep within. Beings that lived then across Asia could have been pocketed here.

The world that presents to people emphasizes sight. But the jungle opens through smell and sound, both of which travel around corners, where sight cannot. While sight needs references, and informs by assigning place, smell functions by chemicals carried to receptors in the nose; smells ride on the medium of air. Sounds also need carriers (air or liquids) but carriage occurs through compressing the medium into waves. The carrying role of media allows smell and sound to travel around corners, wherever the medium might extend. In the jungle, especially when the option of sight is faint as in the dark or abrupt because of vegetation, multisensory understanding is very helpful—for night is the time and deep grasses are the place of carnivores.

Smells feed our knowing through breathing which we always do. Sounds enter consciousness when sleeping, for ears do not have eyelids. With both smells and sounds, animals are able to awake informed. Sight and sound have spoken about what is happening. And beings that live with the wild are aware even when asleep, using senses that citified people may put to sleep.

Here in the jungle we need to know where the animal we seek might be. Lendoop's sketch in the now distant sand gave the picture; we must now etch our place for the hunt. To find a lookout, surveying from the tallest tree is abandoned. Climbing these mammoths is impossible—and even if we could, all trees in competition for light top out at basically the same height. If we can find lookouts, we can see particular features to explore on foot; we can view across the valley to the warm south-facing slopes as animals come into the sun on wintry days. The only lookout possible will be from the ridgetop above.

At the camp's edge stands a maze of shrubs and bamboo. As the morning light starts filling the jungle, Myang, Nuru, Nick, and I begin the ascent to the ridge; it will be 3,000 feet of climbing. Myang swings his kukri through the shrubs and bamboo. Lendoop has left for Shyakshila, as the porters wouldn't return without him. Myang, being barefoot, will turn back with Nuru when the snow gets too deep. A kukri master shouldn't then be as necessary for Nick and me.

At 9,200 feet the bamboo ends and the jungle turns to oak and maple, nearly 100-foot high trees. Little light penetrates to nurture young plants, so walking is easier. Climbing higher, oaks thin out and birches come in. The jungle turns dense again, a mix of birches and bamboo. Does temperature cause these marked vegetation zones, or layers in the sedimentary soil underneath? The bamboo here is thick like jumbo kindergarten pencils and only 6 feet tall. Climbing higher, the bamboo is again high and an inch thick, but now a tight leaf sheath wraps the stalks, and stalks pack close to others. Do the tight-wrapped leaves protect against the cold?

A network of animal trails runs through this bamboo, tunnels in various directions. In one are red panda droppings, in another are serow's. The tunnels are too large to have been made by pandas. Previously we were in the jungle with no obvious trails, why are there so many in this bamboo? All the stubby bamboo shoots may be the explanation—they give a food source and explain this trellis of trails. On a side trail I see different droppings, maybe a musk deer's.

While colonial hunters made expeditions into the Himalaya to collect mounted heads to dress manor houses across England, little was gathered on the behaviours of the animals hunted. The musk deer is important to understand for it might soon be gone, a deer that looks as if it were the offspring of Winnie-the-Pooh's Roo and Eeyore. Nepal's Bijaya Kattel is studying this animal. Here in the Barun, the musk deer is mouse-brown (the colour varies over its range), standing 3-feet high and weighing 30 pounds. It has larger hind quarters than front with a dramatically arched back, giving it its Eeyore-like look. Unusual among deer, this animal has no antlers or horns. Instead, both males and females have tusks 2–5 inches long. With them, the deer scratches while looking for shoots, tender leaves, moss, and grass; and when the snow covers the ground it eats twigs and lichen for this deer can climb out on to the branches of trees.

The larger world would have ignored this deer (as it ignores other Himalayan ungulates that lack massive antlers, such as the barking deer, ghoral, serow, and thar) except for an abdominal pod at the rear of the male's belly that secretes a waxy, aromatic musk. Perfumers prize this pod, as do Eastern medicine-makers, bringing USD 50 in local contraband markets and ten times that in Hong Kong or Paris.

The cute deer is reminiscent of Eeyore in personality, too, comfortably approaching people in locales where it is not hunted, holding to diminutive home ranges of less than 20 acres. When close to one another, males sneeze to warn their ladies of possible danger, protecting their territory through heroic fights during the winter mating season.

The killing method poachers use has thin vines slip-knotted and strung into a lasso. The loop end lies on the ground, the other end attached to a bent sapling. If snared, the light deer is yanked into the air by its feet, dangling until it dies or has its throat cut, poached so humans can smell more attractive—especially sad when one remembers that snares do not distinguish between males, females, young, or old as the kill brings profit only with males over two years of age who have the musk pod. Thus, three musk deer usually die for a poacher to claim one harvestable pod.

☙☙

TREES RULE THE BARUN. Typical pristine habitats have a four-level life pyramid: (a) primary producers, life growing directly from photosynthesis; (b) primary consumers, life forms eating the primary producers; (c) secondary consumers, typically a large herbivore; and (d) tertiary consumers, typically carnivores eating the herbivores. The fact that trees in the Barun are the primary producers creates unusual features, for usually grasses are the base of the life pyramid.

In the Barun, trees are so dense that almost no grass grows as the tree canopy blocks almost all light. Absence of grasses is the reason why the Barun is not grazed by herders, for there is little for sheep and goats to eat, let alone cows. The herders who come are going through the jungle with their herds, headed to the high ridge which Nick and I now seek. Grass, what attracts people first to enter jungles, is not here. The trees have kept out the grasses, and the absence of grasses has kept out people.

When a tree falls in the Barun, perhaps because of age, into the opened space made by its falling rush other life forms—the primary consumers, fungi, moss, insects, and small plants—to rot the trees. Life grown by the tree's photosynthesis becomes food for others. In the usual life pyramid, the size of the organism gets larger with each stage. But in the Barun large trees are the base and the first level of consumers are tiny fungi, moss, and insects. Decomposing together, this becomes the distinctive foot-deep biomass sponge, a reservoir of calories and water. Some primary consumers will be eaten by the secondary consumers: mice whose food is vegetation or voles whose food is earthworms that eat decomposed vegetation. The pyramid continues with larger secondary consumers: birds; red pandas; squirrels—terrestrial and flying—monkeys; ungulates such as ghorals, serows, and barking deer; and the final level, leopards and bears.

Consider the overlooked shrews that live here: they are not rodents and are different from mice and voles. Shrew brains are 10 per cent of their total body weight—no other animal (including humans) has such a high proportion of brain per body size. The shrew here is venomous; its bite can kill a mouse. Some shrews, but not the species here, use echolocation, like bats, to tell them where potential prey might be. And male and female shrews absolutely do not get along (except that moment when they have to). Let your imagination grow shrews so much in size that they become abominable!

With so little grass, the Barun, while having all the expected ungulate species, has few numbers of each. Thus, with a low herbivore population, there are few carnivores. Leopards there are, all three varieties (spotted, snow, and clouded), but very few. Tigers are no longer present as they need vast home ranges, and as people populated the adjoining valleys, the Barun could not support a tiger population. Wolves and foxes are not found because these canines are not forest species.

If there are few carnivores, it means a Yeti that was a true hominoid would have few predators; another reason to be in the Barun with its tree and bamboo concentration: the ideal habitat for a wild, shy hominoid. Here would be less danger from predators. Maybe the Yeti as a hominoid exists? The habitat is similar to that of the giant panda and the gorilla. Those habitats also have few giant predators like tigers and lions that live outside the dense jungle. Pandas and gorillas are both secondary consumers, adept climbers of trees, eating leaves, and some thriving on bamboo.

If the Yeti is like these, the Barun is perfect. Dense trees and steep slopes will cover their whereabouts when moving, allowing them to harvest up and down from whatever mountain vegetation is in season, picking their preference that day. The Barun, like the montane forests of western China with the giant panda and central Africa with the gorilla, offers habitat where in a few miles are found seasonal homes from the subtropics to alpine; and if Nature calls or humans invade, they can always slip over a ridge, cross a snowfield, and gain hiding on the other side.

༄

WE CLIMB HIGHER. THE SNOW IS 1-FOOT DEEP. Thirty-foot tall rhododendrons mix with smaller bamboo and birches. The morning that began in the warm temperate zone has climbed into the snow. Myang and Nuru turn back towards the camp 2,000 feet below. For Nick and me, as we continue climb-ing, the snow gets past our knees, soaking us from the waists down, squishy puddles in our boots. Climb, breathe, sweat, breathe; step follows step, each feeling as though it's lifting a concrete block. A wind blows off Makalu, that great summit still three vertical miles higher. We were acclimatized days ago to 3,000 feet, now nearing 10,000 feet our blood cells search the lungs for oxygen.

Breaths sucking on empty, we climb leaning into the snow, breaking down a trail, headed to the open ridge from where we hope to look out; it feels as though we're getting close. With the slope angle maybe 40 degrees, bamboo and rhododendron coming out of the snow provide a grip, so hands help pull us up. It is almost one o' clock. We left the camp at dawn at six o' clock.

'Hey, Nick, I need to eat something.' We both carry five cha-pattis smothered that morning with peanut butter and jam. To get out of the snow Nick climbs into a rhododendron tree with comfortable-looking branches. I join in a nearby fork. It's great to be free of the wet snow even if there's added wind this little bit higher. Chapatti, jam, and peanut butter melt in my mouth.

Nick, finishing his chapattis, drops into the snow and starts climbing. I pull out another chapatti. Does this snow have any redeeming quality? Maybe the theory proposed by Nepali bota-nist Tirtha Shrestha can explain. He advances that this belt of bamboo and rhododendron, never losing its leaf cover, is a pre-monsoon water reservoir for crops to start in the spring. Across the Himalaya, this interlocking mat from November through March anchors billions of tons of winter snow. The evergreen canopy, as a giant shade umbrella, slows the melting of snow. As heat increases during April and May, climbing these ridges from the plains of

India, the snow melts, seeping into the ground, stoking springs to keep the land below moist.

Shrestha's trickle-down theory of Himalayan ecology holds that as expanding numbers of people go higher on the mountains, seeking firewood or land for corn and millet (the grains that best grow at this elevation), this rhododendron/bamboo belt will disappear, and not only will it result in deforestation, but also in the loss of this water-giving reservoir. The vegetation having been removed, when the monsoon comes in June, the rushing water will carry off the topsoil. In a few years also, with water not seeping into the mountain, the springs will disappear, and the slopes, too, may even start sliding themselves.

Today, though, I find it hard to value this snow. With snow this deep, animals won't be moving. Should I tackle a fourth chapatti? My water bottle has little left in it. Is there enough to wash down that last chapatti and have water to make a plaster-of-Paris cast should we find a footprint?

Nick yells from above, 'Dan'l, there's something here *you* need to see.'

I drop into the snow. No longer aware of having to suck for air, exhilarated at once at a higher altitude, and climbing without oxygen, I crested a ridge and a buddy dropped an oxygen mask over my face flowing at six litres per minute. Nick points to tracks beside my feet. I see the human-like thumb. Tracks march up the ridge from the top of a moss-covered cliff, regular left–right–left prints, like the monogram of a barefooted person walking on two feet. I pull out my tape measure. The stride is 28 inches, the length of the footprint 7 inches.

'Can it be anything else?' Nick whispers.

'See that stride. What two-legged animal walks these jungles? Prints like these couldn't be from a four-footed animal; overprinting of its hind tracks on front would sometimes miss and we would see the other feet. See how it stepped over that branch, yet the print is still perfect. That's what I've always wondered reading

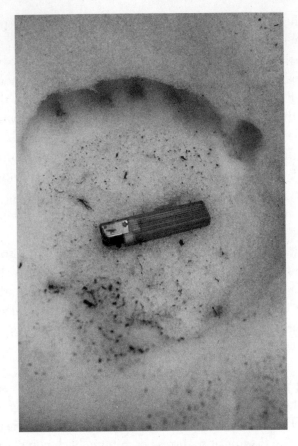

2.3   The Mysterious Footprint Discovered on the Ridge

Source: Author

other accounts—how does the line of tracks look—not just one print. Here is a trail, all prints are clean, sharp. It's bipedal.'

'These are too perfect. The thumb-like mark on each,' Nick says. 'You don't just stumble on to a discovery like this the first day out.'

'What do you mean by "too perfect"? You want to find something that biologically doesn't work? You want those turned-around backward feet always mentioned with forward walking? What's wrong with discovery the first day if we're looking in the right place and have been trying for twenty-seven years? Why did I forget the plaster of Paris today?' Disgusted after rooting through my backpack, gear now across the snow, I turn to taking pictures.

The moss on the cliff is slightly disturbed. Bamboo grows by its rock face. On the base of some stalks are scratch marks. At the top of the cliff, broken stalks hang over the rock, appearing like these stalks were used to pull the animal up. The mossy wall shows where feet were placed during the ascent. A fastidious climber went up this face, an animal with balance and considerable strength.

Kneeling in the snow, Nick and I examine the tracks, seemingly consistently left–right–left, each with the thumb-like print, and some prints a lot sharper.

'Do you see any evidence of any track of overprinting by hind feet coming down on top of fore?' I ask, looking at Nick.

Tracks move from the left side of the ridge to the right as they go up the hill, then to the centre, back to the right. As we follow them up the ridge, sinking into the snow, it seems the maker, apparently passing earlier that morning, walked on top of the snow without sinking. The animal somehow also sensed where branches were below the snow, branches that kept it from sinking through, using these branches like in-place snowshoes. Every so often a foot falls through.

Most tracks are less than an inch deep. Puzzling black dirt in some tracks is probably bark off the branches above. Tracks under trees are sharper than those in the open. My watch says 4:22. We spent eight hours climbing and now two hours up here; there are a couple hours of daylight left to race back to the camp. Tomorrow we'll see how these tracks have changed. A trail has broken open now to the ridge.

three

The Bear Mystery

Previous page:
3.1 The Gorge into the Yeti's Jungle

Source: Author

'Papa, what are we going to do?' Jesse asks, finishing a breakfast of rice cooked in milk. 'It's raining. I'm cold.' He and I stand under the plastic tarp by the cook fire. Jennifer's gone high with Nick today to check the tracks. Off our elbows drips the now-familiar Barun drizzle.

Back in the tent we read *Winnie-the-Pooh*, discussing the challenges of searching for honey, wondering if anything else might hide in the 'hundred-acre wood'. A break in the rain gives time to leave the tent and search 'for Owl's house'—Jesse reminding me that we heard an owl last night. Outside, we build Heffalump traps in our wet thicket, doing that makes us numb, so we go back in the tent to hang the laundry where body heat may slowly dry it. Then, to the woods again and we rig a climbing rope from a branch which lets us pirouette tree-to-tree in an imagined Yeti style.

Nature's most-tested classroom has come alive. Where interest leads and surprise lands opens opportunities for father–son lessons—they are caught not taught in questions that opened in his mind. We search under rotting logs, take apart a tree cavity for beetles, and develop our taxonomy: beetles with horns, beetles that look blue, beetles with pincers that hurt, beetles with wings allowing them to fly away. Life is discovered.

'They're almost melted, Dan'l,' Jennifer says as she and Nick return. And setting down his pack Nick adds, 'Even our tracks from yesterday. If our tracks melted so, then for the tracks to be as sharp as they were yesterday, they must have been made only hours before.'

'It was good to see your tracks, but really much wasn't there.' Jennifer looks up from the stool on to which she plopped, taking the mug of sweet tea and two Fig Newtons offered by Pasang. 'The nicest part was catching the mood of that high jungle. But your tracks aren't convincing.'

I turn to Nick. 'How long before us yesterday do you think the tracks were made? It would be about the same as for Shipton who also came in the afternoon.'

'It's amazing, Dan'l, how fast tracks melt. The tracks we saw were sharp, the toes clear. Now they're snow bowls despite no sun today. The only prior aspect still sensed was right–left bipedalism. Tracks when we found them had to be fresh, but if they were that fresh how did the bark flakes fall in so quickly. That "something" must have been shaking branches as it walked.'

Holding Jesse, Jennifer says, 'We followed the trail up the ridge until it dropped off into the bamboo. While following, I stepped on a branch on top of the snow, a branch the animal had also stepped on. The branch bent under my 110 pounds. When Nick stepped on it, the branch cracked, actually splintered. He's 150 pounds, maybe 170 with his pack.'

'So this beast is less than 170 pounds,' Nick says with pride. 'We started looking at other branches. We found a branch that crossed the path. A four-legged animal should be able to walk under, but a two-legged animal would find the branch blocking its path. So how did our animal go? There were no marks in the snow of four feet, so it doesn't seem to have dropped to all fours. But on the back of the branch was a scratch, a scratch in the moss on the top seeming to suggest the animal swung under.'

'They were scratches like those on the bamboo on the cliff,' adds Jennifer.

That night soft flakes of snow started, no ting-ting as with rain, or click-click like dry snow. Awakening, I realized the tent walls were pressed down and might collapse the poles. Going out to scoop snow from the walls, night opened; in a world where sight was turned off, sound and smell gave the jungle fullness, a valley the king had called 'the wildest in my country'. Here, surrounded by the highest of mountains was a true wild, a place more wild than the 4,000-times-visited Everest summit just miles away. Might it be that people who go into such wilds are discovering the Yeti in themselves?

❧

PIGLET, TOO, HAD TO BRUSH AWAY SNOW, Jesse and I discussed the next morning. And as Piglet was doing that, he looked up and there was Winnie-the-Pooh walking round and round, thinking of something else. And when Piglet called, Pooh went on walking.

'Hallo!' said Piglet, 'What are *you* doing?'

'Hunting,' said Pooh.

'Hunting what?'

'Tracking something,' said Winnie-the-Pooh very mysteriously.

'Tracking what?' said Piglet, coming closer.

'That's just what I ask myself. I ask myself, "What?"'

'What do you think you'll answer?'

'I shall have to wait until I catch up with it,' said Winnie-the-Pooh. 'Now, look there.' He pointed to the ground in front of him. 'What do you see there?'

'Tracks,' said Piglet. 'Paw marks.' He gave a little squeak of excitement. 'Oh, Pooh! Do you think it's a-a-a Woozle?'

'It may be,' said Pooh. 'Sometimes it is, and sometimes it isn't. You never can tell with paw marks.'

With these few words he went on tracking, and Piglet, after watching him for a minute or two, ran after him. Winnie-the-Pooh had come to a sudden stop and was bending over the tracks in a puzzled sort of way.

'What's the matter?' asked Piglet.

'It's a very funny thing,' said Bear, 'but those seem to be *two* animals now. This—whatever it is—and the two of them are now proceeding in company. Would you mind coming with me, Piglet, in case they turn out to be Hostile Animals?'

໑∾

NICK AND I LEAVE THE CAMP THE NEXT DAY in different directions. Returning that evening, neither of us has discovered much. Animals are not moving with the new snow, and back in the camp we discover we've lost the possibility of dry camp slippers as the day's melt has created slush that swamps over slipper tops. The next day more snow falls, so we cannot ascend to the ridge. It's easy to justify taking a day off. Towards noon the snow turns to a drizzle, and in the evening back to wet snow.

3.2   Our Camp in Makalu Jungli Hot'l

Source: Author

Suddenly shouts come up the slope. It is Bob! Lendoop and Lhakpa bring him in. Huddling together under the plastic tarp with mugs of hot tea, Bob probes, 'Did you see nail marks in the snow? Are you sure? Then how did you find nail marks on the bamboo, on the cliff, even on the branch? Are you suggesting it's an animal which can have nails out while on cliffs and holding branches, and like a cat draws them in while walking—if so, they are claws not nails. More importantly, it's not a primate, certainly not a hominoid, or even a bear or panda.'

Dampening my earlier certitude now are questions. Eight years older than me, for my jungle-learning Bob Fleming was a mentor. As a four-year-old, I tagged along as he went out with his gun and brought home bird specimens. Aside from knowing animals and plants, he's also mastered several Himalayan languages.

Bob goes on, 'Why are you sure there are no hind-foot over-prints? Did you take enough photographs? From your description, you were more than exhausted. What? You found the long-sought tracks and didn't use a whole roll of film—then back that up using another full roll in case something was wrong with the first roll? And, dismiss the idea of the animal shaking branches and bark falling in the steps—wind did that, you're on a Himalayan ridge. What real evidence do you have that the route showed intentional selection for branches under the snow? Did you pho-tograph that? What the Yeti has always lacked is evidence, not hypotheses.'

'Bob, we have mysterious footprints—they fit in size with Cronin and McNeely's, also Tombazi's.'

'And again,' Bob replies, his voice holding genuine curiosity, 'we do not have any evidence beyond the footprints as to what made these mysteries.'

The next morning begins our sixth day in the camp. Freezing rain leaves a film of ice on rocks and a crust that with every step breaks to slush beneath. Bob looks for a project to sustain his optimism, and, practical as always, suggests that the probability of breaking legs will lower if exploration is done by talking. So we retreat into the large tent as he questions Myang, Lendoop, and Lhakpa.

He starts by joking with the three about animal sex life. Bear with goat—what would that make? Monkey with rat? The jokes give a list of jungle animals, similar to mine of almost two weeks ago at the confluence, but with three new ones, the wild dog and both species of civets. Bob slips in oblique references to the Shockpa, passing a snippet from one account he'd heard with a piece from another village.

Then Lendoop replied, 'I am many times in the jungle as a hunter. How can there be such an animal? I never see its sign in the jungle. What does it eat? How does it move? I have heard shockpa stories from people but I never see any signs myself.'

Bob casually asks, 'Lendoop, if not the shockpa, what did Dan'l and Nick find on the ridge?'

'I didn't see the tracks. Maybe *rukh balu*.'

'Rukh balu?' Bob queries offhandedly.

'There are two bears in this jungle. One is *bhui balu*. It is black, strong, very aggressive, and when dead requires five men to carry out. The other is rukh balu. It is also black, but moves in trees like a monkey, and this bear is shy. Two men can carry a dead tree bear.'

Bob speaks carefully, 'One bear lives on the ground, another bear in the trees, different sizes, living in different places. One aggressive, one is shy. Are there other ways they differ?'

'It is strange,' Lendoop answers. 'The front paw of the rukh balu is like the human hand. I think the two sahibs saw that in the snow. A tree bear can hold things with the thumb on one side and other claws on the other side.' Lendoop shows with his hands an opposable grip between the thumb and fingers.

Bob and I are quiet. A bear in a tree with a forepaw that makes prints like a human hand. Searchers have been looking on the ground because footprints on the ground defined the

3.3   Bob (extreme right) Interviewing the Team in the Tent

Source: Author

beast—while the animal that may make those prints probably was seen many times, but in trees far away to not be recognized as different but so close in resemblance that its identity is mistaken. A small black bear—rukh balu—shyly hiding in trees while explorers and scientists pass underneath. If they saw the little bear they would count it as a cub of *Selenarctos thibetanus*, a known species on their taxonomical list. They were looking for a man, an Abominable Snowman.

Lendoop goes on. 'We see the rukh balu when it comes to our fields to eat corn at the end of summer. In Shyakshila I have a skull of the animal I killed five months ago.'

Lhakpa breaks in, saying that in his grain room is a tree bear skull along with its dried front paw and back paw. He keeps them to scare away mice and evil spirits.

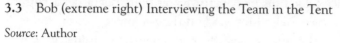

WE SIT BY THE COOK FIRE. What do the villagers believe the Yeti to be? Those sitting beside us do not know the Yeti as 'Yeti'; they do not even use the name that is nearly universal throughout Nepal of *bun manchi*, 'forest man' (but they are aware of the name as Lendoop showed when explaining the footprint in the stone). Their mysterious jungle men are the Shockpa and the po gamo, as evidenced by the mountain's name while entering the Barun Valley.

The po gamo appears to be a spirit that takes a physical form. It attacks people at night who walk a trail, frightens homes when people are inside, and kills domestic animals. That it must be a spirit is proven because both knife and gun are useless in defence. Protection comes from substances holding cosmic power, perhaps gunpowder folded up in paper; Myang pulls out a dead battery, saying its power helps him ward off the po gamo.

Almost all cultures have deeply believed in animations hiding on the edge of the spirit world; familiar names for these are ghosts, angels, or spectres. In such usage, ancient soundings from our human souls have sensed something we do not precisely know as sentient beings; we touch there with sensation but it is beyond cognition. Science may deny their existence, but such belief crosses cultures (only most of these apparitions do not make footprints). As we sat by the fire, Pasang described another beast, the needene.

Quite astounding, this needene—an animal only in the female form—is a lady with pendulous breasts half-a-metre long. When running, these women flip their breasts over their shoulders to stop them from smacking. Finding a child to kidnap, the needene touches the child, which makes the youngster speechless, and then she carries the child to her cave. When a needene leaves to search for food or more children, the dumbness departs and the youth then wails plaintively; this their parents hear in the valleys below. To these captured ones the needene brings insects. When searching for such food, if the needene does not return soon, the children wander, and their parents may find them stumbling among the rhododendrons. Recovery from the dumbness will happen when the children are slowly fed ash soup.

It was clear that these realities are believed, and in the reality in which they are held the line between physical and supernatural existence blurs. Physical reality and supernatural cognition allow a spirit to dwell alternately, simultaneously, or in only one state. Perhaps, we thought as we talked by the fire, this is because their Buddhist view emphasizes that life is in a world of imperfect understanding where things that appear concrete can be illusion, and illusions mask the identity of the real.

When absorbing life by a campfire in the jungle, the world of streets, timetables, and cities must be relegated to one orientation. There in the jungle, where life does not naturally linearly connect cause to effect, a different awareness opens. In the jungle, reality is of Nature's making not human, understanding osmosis from existence itself as we listen to a different understanding. Concepts grow, and believability changes. Of this world what most of us know are stories, and though fascinating, pragmatism denies their truth. But a campfire in the jungle ignites such consciousness; learning comes from beyond that which can be measured, beyond sounds and smells. Senses exist that we cannot touch, smell, see, or figure out. Villagers experience events, then explain these realities that have been imprinted in their lives. Something came into their lives from the wild. It is inadequate to say 'It happens', for real life is messier.

In making their lives, almost all cultures choose to live apart from the wild. Even Shyakshila hangs on a slope outside the jungle. Its people may visit the jungle, but they do not live there. Homo sapiens seem to seek a human-made environment. The major feature of habitation that differs between wealthy people and villagers is that villagers build with materials from the land, whereas the wealthy bring in from afar, and do not do the building. The objective is the same: to get away from the wild.

By being around the civilizing influence of a fire, asking about a wild man while still encircled by the jungle, what is discovered is defined by circumstance—so might it be the glacial context that shapes the Yeti observations of tired mountaineers. Our senses and our moods define our objectivity. Similarly, trying to interpret evidence in a world of science presses the Yeti to a definition which is not where the Yeti originated. Finding a footprint, does the lonely mountaineer think *wild man*, or does he think, 'Ahh, *some man was here before me?*' Each individual, whether a villager, or a tired mountaineer, or a scientist, thinks not just in their worlds but also in that particular moment. For villagers, the explanation is simple: a sign of an apparition from the wild, then he or she walks on. The relationship to local life got answered. The intent was never to explain taxonomy.

Is this explanation really true for Nepalis? Bob and I try to remember as we have been walking these mountains for decades. One thing we had noted, that explains the Yeti, is that the now extensive Yeti lore may have become magnified because of Western interest. What happened was Westerners picked up then amplified a feature that Sherpas saw and simply walked on from. Outsiders went into the valleys, and seeing Everest was not enough. Back in Kathmandu, they went to the Yak and Yeti Bar, got yakking, then saw all sorts of wild men. Nepalis discovered money was made off T-shirts that portrayed the Yeti that way. Bob noted how in religious *thangka* paintings the older paintings show demons and spirits, but no Yetis. Today, on the new thangkas, Yetis can be seen peeking from behind Buddhas. Might it be Yetis are being seen now because people want to see them?

Indeed, what do Nepalis see? Again, it is footprints. But footprints found by Nepalis indicate backward feet, toes pointing away from the direction the animal walks. Is there any proof, Nick wondered, of toes spun 180 degrees? Bob reported Sherpas out with their flocks seeing Yetis, and when he pressed such herders their reply was, 'I've not seen myself, but my father saw when he was a boy with yaks high in the pastures.' Tenzing Norgay, who first climbed Everest, told Bob just that. In sum, there is no evidence except always the footprints, and now it is not even clear in which direction they were going.

Nick was fascinated by the flipped feet: physiologically, wondering how connecting the foot up with bones and musculature backwards would control the forward falling that is called walking. Hominoids have two distinctive physical abilities: speech and walking. Both physical abilities have specialized structures. And for walking more is involved than just the feet, paraphernalia from balance hairs inside the ear to kneecaps and complex neural–muscle connects. It would be impossible for reverse evolution to turn around such complexity (unless evolution had gone in two directions many millions of years ago).

But as evidence of cosmic power, the magic of turned-around feet makes the Yeti—that is the villager's point. If the footprint-making Yeti is a bear or a monkey, flipped feet make it a superhero. When baffling things happen, that now have an explanation, it is easy to overlook complex connects from balance canals to toes, or life in a world where there is little food and many challenges. To these, heroic answers, like religion, give explanation on the trail of life meaning, answers every bit as real as science, just evolving from a different reality.

That 'Yeti' is a uniquely Sherpa word is relevant. The Tibetan language from which the Sherpa language descends does not use *Yeti*, but uses *me tay* for this beast in the same manner as the Nepali language uses bun manchi. A me tay in Tibet can outrun a human; it lives in rocky mountains (a puzzling habitat as a rocky terrain doesn't have much food and lets a reclusive animal be seen). And the me tay grows even less logical as real since one of its favourite foods is reportedly frogs, and above 10,000 feet amphibians are rare in the Himalaya, especially in rocks. Anyway, when crossing a stream and seeing mud and disturbance, lore says to suspect this indicates a me tay is frog-rootling upstream.

A me tay can grow 10-feet tall. It can come into a yak herd, sling a 200-pound yearling yak over their shoulder, and walk off. Footprints are never found, and when a me tay eats these lost yaks no remains can be found. To this Nick suggested, 'The me tay might be a story developed by yak poachers. They creep up during the night, make tracks that appear to be coming when actually going, grab a yak, maybe dye the coat to mask its identity; suddenly a group of poachers is one yak richer.'

But a logical explanation is not what the yak herder wants. For a herder caring for someone else's yaks, his problem is that one yak has disappeared. The likely explanation is that the yak, while the herder was inattentive, walked off and fell over a cliff. But to avoid being accused of carelessness, the herder who must report to the yak owner, gives an explanation that cannot be followed up. Is the owner going to search the mountains for that yak?

With a mystery, does one look at the footprint, or the trail? It is easy to focus on the footprints when the story lies in the trail. For a Nepali, what the Yeti provides is not a mystery but an explanation. To the scientist who sees flaws among the facts, the Yeti is a masquerade. But across cultures and across a century, for some the footprints go forward, backwards for others, and whatever direction the prints signature on a mountain slope, they imprint meaning.

In one way, it is like sunrise across the same slope that also speaks of undeniable magic that transcends what is happening physically. Does a viewer really want to think about the refraction of light while looking at a sunrise? Yet the footprints that keep returning, for the scientifically inclined, call for answers of taxonomy. For the spiritual, what had been given is an animation of life-force worlds. Still, while this discourse goes on and on, with the Yeti always there and undeniably real are footprints. Because no one sees that particular set of prints being made, what is seen becomes defined by what one wants to be seen.

ᵔᴥᵔ

THE NEXT MORNING BOB LEAVES to return to the confluence where his wife Linda waits. As he leaves, Bob questions how sure I am that the ridge tracks had no nail marks. For bears, and certainly tree-climbing bears, have nails; and bears do not walk on two feet. He points out that bears, like many animals such as house cats, place their hind feet precisely into their forefoot prints, making the prints look bipedal.

That morning Nick and Jennifer prepare to leave the camp to implement what we all know is a crackpot scheme, for we've concluded we need to call in our animal and watch it make prints. But, never having made a Yeti lure, nor having successful traps to compare to, our thought is to cause a tree smell. Draw the mysterious footprint-maker, presumably now a tree bear, bait it, and have snow around the tree able to take the prints. We plan to scatter scents in the air. The enticement will be spreading peanut butter on branches and inside hanging cans punctured with holes to let the peanut smell waft out. Bears on four feet will be looking on the ground, and those on two feet will be reaching up. Bears that can climb trees will be going up. Years before while guiding in the Grand Teton Mountains, as I was reading a book, a black bear came inside my tent and on top of me because it smelled the peanuts I was eating from a can beside my head.

Jennifer and Nick will distribute our peanut-flavoured traps in the bamboo belt that is plentiful with bamboo shoots. The wind will waft peanut aroma around the ridges. Grandpa reported that a Himalayan black bear could smell a rotting animal 15 miles away. Tomorrow we will return to the trees and see what has come. As they go off, staying in the camp, Jesse and I turn to stories.

⟲⟳

WINNIE-THE-POOH SAT DOWN AT THE FOOT OF THE TREE, put head between his paws and began to think.

First of all he said to himself: 'That buzzing-noise means something. You don't get a buzzing-noise like that, just buzzing and buzzing, without its meaning something. If there's a buzzing-noise, somebody's making a buzzing-noise, and the only reason for making a buzzing-noise that I know of is because you're a bee.'

Then he thought another long time, and said: 'And the only reason for being a bee that I know is making honey.'

And then he got up, and said: 'And the only reason for making honey is so I can eat it.' So he began to climb the tree.

૭૦

THE FOLLOWING DAY NICK AND I DEPART in separate directions. My task is to search the next valley for signs. Nick will check the cans. On an east-facing slope I find a fresh leopard scat filled with brown hairs and big bone chips. Was it a common leopard, or maybe the rarer clouded leopard that lives in dense, warm temperate jungles? The altitude is too low for the snow leopard. From amid a large stand of maples, a ghoral stares at me, peeking over a fallen tree, and then bounces away. With just its head visible and the light on it shadowed, was it really a ghoral or a juvenile serow? The photograph I squeezed off will tell.

Entering the belt of thick bamboo, crawling through the two-foot-high tunnels, I come upon fresh red panda scat. Ahead, the hind end of another animal disappears. White legs … a serow? Or did the colour change by a flash of light? Or was it a musk deer? Suddenly I wonder what I should do if a bear charges in this bamboo tube. I pull out the canister of tear gas mixed with the oil of pepper. No one has used this on Himalayan bears, but some folks in Yellowstone now carry it. I open my water bottle, moisten my bandana, and pull it around my neck, ready to draw up over my nose and mouth in case I must shoot the canister.

Breaking out from the standing bamboo into where wind or snows have matted down the once 20-foot-high bamboo, the stalks are packed and vines entwine with the bamboo, a total thicket. Lacking a kukri to cut my way and remain on the ground, I slither to the top of the mat. I have no idea what mashed this bamboo down. As I step out, the mat heaves like a trampoline. Two-thirds of the way across, my feet slip, zipping me down. Suspended by my backpack straps, I am caught above the ground.

If I try to get out by falling to the ground and forcing my way, my pack will remain hanging in the bamboo above. My right arm inches into my pocket where I have parachute cord, working that hand back until the cord is in the front of my teeth. Inches from my face, framed by bamboo green, is a beautiful red, black, and yellow diamond weave. Using hand and teeth, I tie the cord into a bowline. First, the flip of a small loop, then out of the loop comes the rabbit (as we said in Boy Scouts), around the tree, and it ducks back in. I pull the rabbit tight. That loop is dropped to my boot as a stirrup.

Now knot the other end—bend it into a bight, then loop that around the big bamboo near my face, feeding the string through the loop. Do it again, then again: a Prusik knot. Swing the stirrup end until it loops over my boot, easing weight on to that. The beauty of a Prusik knot is that it slides up a rope (or bamboo stalk) when weight is off but locks when weight is on. I wedge myself and unseat the knot so it slides—one step and I am 4 inches higher. Lifting my foot and slackening the cord again, 4 inches more and shortly I'm back on top of the bamboo mat.

I pull out a squashed chapatti from my pack, and look up at the sky; a lammergeyer eagle circles, its long tail identifiable it as it rolls and turns. The biggest bird in these mountains, it's a vulture, really. They soar with a 9-foot wingspan on currents of ridges, wings flat like the griffon's, not folding into the 'V' of a golden eagle. This bird is probably looking for carrion. Up tips one wing—and the great bird rolls. Up on the other wing—it rolls the other way, effortlessly like a pilot in a Pitts Special acrobatic plane. Though the undersides of the griffon and the lammergeyer are both white, this eagle, when doing its splendid turns, can never be mistaken for a griffon. Aggressive, the lammergeyer sometimes chases the griffon whose 8-foot wingspan is only a foot less. Seemingly more creative than the griffon, finding carrion, a lammergeyer drops the bones on a rock, and from the shattered fragments picks out the marrow. As I watch, another lammergeyer comes into view. Does circling by the first call this second, perhaps sending word of a body lying motionless on the bamboo below—but, no, what I see is courtship. Down swoops the first bird, showing his prowess before her whom he has attracted; straight down, then with a dip of that long tail, shoots up, over the top, making an inside loop more smoothly than any pilot in a Pitts Special.

The doggerel, that our favourite spinster, Miss Marley, teaching English to us at Woodstock School in the Mussoorie Hills, comes back to me:

> What is the mind of the vulture ...
> That sits and thinks
> That stares and stinks
> And has no culture?

Anyone who watched this bird could never assume such a soaring creature lacked culture; indeed all vultures (a bird that never kills but removes the kill of others) in attending to their ecological roles are efficiencies of flight and immaculate in their grooming. When writing her poem, Miss Marley might well have listened more closely to her pupil, Bobby Fleming. For it was Bob who explained to me the behaviour of this greatest of Himalayan birds. As I watch the couple above dance, twisting together in flight, with a sudden chandelle turn, one bird takes off, the other following over the ridge, neither now a spinster.

Over supper Nick reports how he lost the trail in the fresh snow that was at higher elevations, so he couldn't find the cans and has no idea whether 'something' found them. But in a clump of bamboo he found maybe forty stalks broken at a 3-foot height. The

stalks were an inch in diameter, hefty bamboo, that by himself
he could not break. Yet these were snapped once, then again in
another direction, to make what looked like a nest.

The next day Nick and I together try to break that bamboo.
The animal that made this 5-foot-wide nest seems to have snapped
stalks wherever it wanted. One clump of bamboo was shaped into
a basket. Is this the handiwork of a tree bear seeking to make a
sleeping place above the soaking snow? I am familiar with birds
making nests, even squirrel nests, but here a very big animal is
evidencing basket-like handiwork.

Almost 6 feet above the first is a second nest, also made of broken bamboo. The animal seemed to have sat in the nearby rhododendron tree, reached out and brought bamboo over and bent the
stalks around a fork in the tree. Ten feet away is a third nest, smaller
than the first, made similarly. Is the bigger nest the mother's and the
two smaller up higher for offspring? Did each animal make its own,
or did the mother (if that is the explanation) make all three? Did
they stay one night, or did they use these nests for several nights?

We search the nests for hairs. But snow has been falling for
half-an-hour. Nick's lips are blue, and he's having trouble talking. I force him to drink the rest of my orange drink, still warm
from being rolled up inside a down vest in my pack. He perks up
and talks clearly. I also give him the vest as well as a dry ski hat;
somewhat normal now, he heads back to camp eating our last two
candy bars. With hypothermia, the important thing is to identify
it before body temperature starts to fall. Our footprints coming
give him an obvious return trail.

Returning to the search, nail marks can be seen on the tree trunk
a little more than a yard above the second nest, with more marks on
the baskets' bamboo stalks. Three stiff black hairs, maybe 2 inches
long, are wedged against tree bark. There are two more hairs in one
of the tree nests, stiff black hairs precisely like those of a bear.

Descending to the camp and now looking into trees, I spot
two limb clusters high in two adjoining oak trees. Both look 5–6

3.4 Nick Looking up at a Bamboo Nest in the Rhododendron Tree

Source: Author

feet across and maybe 60 feet above the ground. Studying them through binoculars, I notice that the branches are stacked much like the bamboo. Up the tree are scratches that slant diagonally. Why are the scratches not straight down, reflecting an angle of pull as the animal climbs or descends? Then I realize this animal did not claw up like a cat. Each oak has a thick vine growing alongside the trunk and on the back of these vines are nail marks. The animal seemed to have grabbed the vine with its front paws, leaned back and pulled, hind paws on the trunk, climbing like a monkey up a tree. The animal I suspect Lendoop calls 'tree bear' certainly acts apelike.

ᄋᚖᕉ

POOH SAT DOWN, DUG HIS FEET INTO THE GROUND, and pushed against Christopher Robin's back, and Christopher Robin pushed hard back against his, and pulled and pulled at his boot until he had got it on.

'And that's that,' said Pooh. 'What do we do next?'

'We are all going on an Expedition,' said Christopher Robin, as he got up and brushed himself. 'Thank you Pooh.'

'Going on an Expotition?' said Pooh eagerly. 'I don't think I've ever been on one of those. Where are we going to on this Expotition?'

'Expedition, silly old Bear. It's got an "x" in it.'

'Oh!' said Pooh. 'I know.' But he didn't really.

'We're going to discover the North Pole.'

'Oh!' said Pooh again. 'What is the North Pole?' he asked.

'It's just a thing you discover,' said Christopher Robin carelessly, not being quite sure of himself.

'Oh! I see,' said Pooh. 'Are bears any good at discovering it?'

ᄋᚖᕉ

'PAPA,' JESSE SAYS, 'I'M ON AN EXPEDITION TOO … just like Christopher Robin.'

We are sitting around the campfire, having just finished supper and I'm reading to Jesse. Lendoop returns having walked Bob out and shown his bear skulls, and he's eager to take us to Shyakshila, saying we can make it in one day and must leave the next morning. We tease him that it's too fast, because it took the porters three days to trek in. Then he flips the rag on us, 'Jesse's mother walks like a Nepali woman—she also carries a heavy load.' I laugh. Given her eagerness to get out of this jungle where it is always raining, she will be doubly strong, for at the Barun's jungle-free confluence she'll be able to sit in the sun near what Jesse calls Poohsticks Bridge.

That last night in Makalu Jungli Hot'l, from my sleeping bag I listen to noises give description to the night. If we stayed another day, could I find the peanut butter cans? Another day, and maybe we could find more nests and the chance to study them carefully? I'll soon be planning another expedition—but through exploring this jungle with Jesse I've lived again my jungle years with Grandpa and Dad. Beeps of that giant flying squirrel come out of the night. When we listen to what we don't see, life outlines our wild world in so many ways.

Jennifer awakens me before dawn, trying to be quiet as she packs. Outside, pots rattle as Pasang and Nuru are at work taking down the cook site. Pasang calls to Lendoop to strike the porter's tent. Before dawn turns into day everything is packed, and by eight breakfast is over. The four porters from Shyakshila haven't arrived, but no one wants to wait, so we pile on doubled loads. Lendoop smiles, looking across as Jennifer tightens what must be a sixty-pound pack.

We start off; when one person slips, another supports. Jokes pass member-to-member down the muddy trail, hands outstretch to each other moving almost as one, creating in our group the feel of a centipede with all its legs. Pasang's basket with another basket on top jangles with the cacophony of spoons and plates nested inside pots inside larger pots. Yetis, bears, and crimson-horned pheasants will hear our out passage as the jungle returns to being theirs.

The four porters wait at Payreene Khola; they had stopped in the opening of the landslide for their morning rice, assuming they needn't hurry in as we would be leaving tomorrow. Steam and smoke from their fire curl together with mist from the waterfalls. Loads are redivided as we drink sweet hot tea. Then as we shoulder our loads, the momentum of home carries us like the current of the Barun from the heart of the jungle.

The next morning, Lendoop and Lhakpa wait outside our tents. I had hoped they would bring their skulls to us, my knees sore from yesterday. But skulls are not their priority; their desire is to host us in their homes. Pasang, having heard so much about Shyakshila while he cooked, wants to see their village too. Our first stop is tea at Lhakpa's, then boiled eggs at Myang's, where Myang's wife cannot take her eyes off Jennifer and Jesse. She blushes when Jennifer speaks to her, and doesn't respond though we had seen her repeatedly at our camp. Her two children come out of the back room, and we pass along several of Jesse's T-shirts.

At Lendoop's house, before we enter he explains the immense boulder in his yard, half the size of his house. The autumn before, land loosened by monsoon rains, this boulder rolled down, stopping only because he had levelled a little terrace there in front of his house. 'My life and my family are lucky,' he says. 'Disasters always miss me.'

Luck is clear: that house-sized boulder would have pancaked his family and house had its path slightly differed. Inside his house, light streams across hewn logs that shine with marks of an adze crafter who knew his skill. Years of padding feet have polished a patina to the wood. As we sit on Lendoop's ground-bear skin, a trophy from one of his hunts, his wife brings us food. Then Lendoop brings out his bear skull. But I've already bought Lhakpa's, for that had front and hind feet as well. One skull should allow us to determine where the tree bear lives taxonomically as a species.

four
My First Yetis

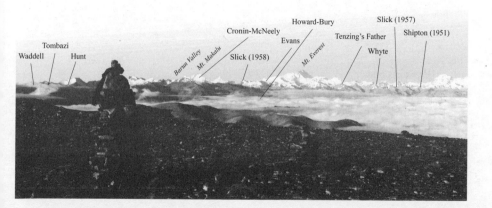

Waddell · Tombazi · Hunt · Barun Valley · Mt. Makalu · Slick (1958) · Cronin-McNeely · Evans · Mt. Everest · Howard-Bury · Tenzing's Father · Whyte · Slick (1957) · Shipton (1951)

J uly 1956. Twenty-seven years before I found those footprints on the Barun ridge, I crossed the Yeti's trail. I was eleven years old. We lived above the town of Mussoorie in the Indian Himalaya, the last bungalow at the top of the hill—it took over an hour to hike there from where the motorable road ended—an expansive compound, and in those days there was the jungle all around with towering deodar cedars.

My grandfather purchased this mountain home in 1920 for a hard-to-imagine low price in an India hungry for cash after the First World War, a land having also just endured the Great Influenza epidemic. With a population in 1920 one-fifth that of today's India and one-third of that in 1956, an India then alive with jungles, it was an India that taught different skills to a child. On waking up every morning, I had to think about a snake maybe having come in through the bathroom drain, and to remember to shake out possible scorpions from my shoe.

Walking into the jungle was like stepping out the back door, a skill of everyday life—in the way a child in the city learns to cross wild traffic when stepping off the curb, or a child on a farm learns to climb aboard a moving tractor, or children everywhere learn to navigate parental tantrums. Whatever world they grow up in, for a child that world is normal. The idea that this normal might be unusual is one that comes as they enter new worlds, just the way one learns that one's native language is not spoken by everyone.

That day the monsoon was pouring outside our bungalow. While Grandma took a nap, I quietly crept into the kitchen. A screened-in cabinet was where cakes were kept for teatime, and my prey was a fold of icing; a careful swipe of an eleven-year-old finger could remove it undetected. The adventure of the steal lay as much in the thrill as in the sugary bite. But while passing the dining-room table I was stopped by a photograph on the newspaper, a footprint. At the age of eleven I met the Yeti for the first time.

Rain shook the windows. For two generations, grandparents, uncles, aunts, and cousins had come to this hill station to escape the summer of India. Habits begun in British times were changing now, from the way we dressed (as in the front hall, hanging off deer antlers were our pith helmets seldom used any more) to the way we looked at the jungle (wondering whether it was right to kill animals as we once easily did). India was changing, as we had only years before lived through the birth of the newly independent India.

For a 'white' child growing up in those optimistic days of the framing of a new country, melding relics of our history with the changes all around, my life at least still circled around the jungle. A python skin curled above the trim down the hall, a poster also hung nearby showing the different types of mosquitoes as well as the life cycle of malaria-making mosquito, and walking staffs to accompany one's walks stood ready in a corner by the door. Stories unfolded with all these, often told from the wide swing on the veranda (from where Aunt Margaret as a teenager had shot a leopard twenty years before), a swing that looked out into those splendid deodar cedars. Artifacts and stories, all were parts of a large foreign family's trying to understand in an India which we had also made into our home.

For three weeks now the monsoon had clattered on the tin roof, clouds had wrapped the house, where with the monsoon the pervasive smell of mildew would last another two months. Today was a Saturday, and usually on Saturdays, without school, I went into the outside world. But trapped today by the monsoon, with no more icing to be gleaned without notice, I took the newspaper to the living room; a tiger skin with a mounted head was draped over the piano, and I settled into an old armchair to read.

The newspaper's picture had been taken on the 1951 Everest Reconnaissance Expedition. That expedition had opened a region until then closed to non-Nepalis. Nepal interested me, as two years before that Everest reconnaissance my father, as a medical doctor, had been on the first expedition of foreigners to enter central Nepal, then a land more unknown than Tibet. In exploring the Everest region, Eric Shipton and Michael Ward found this footprint high on the Menlung Glacier.

And now in 1956, the *Statesman* newspaper was saying that other mysterious human-like footprints had been discovered. The article asserted that independently taken photographs five years apart meant that the Yeti moved from a possible freak into fact. Photographs across such a time made it unlikely that the 1951 prints were from an abnormal individual. A mysterious species must be walking across the snows. Legends do not make footprints. The shape was clearly hominoid.

And the foot's size, more than 12 inches long, was near superhuman. With the newly discovered prints which were, the newspaper said, smaller than the first—about 7 inches long—there was not only a mysterious maker, but also a new mystery of size. Were the new prints a juvenile's? Neither Shipton nor Ward, the two mountaineers who had taken the original photograph, made the Yeti claim. They had simply photographed what they found. But with added discoveries, a Yeti claim was being widely proposed.

So, in July 1956, my Yeti quest started.

၆၃

A CURATOR AT THE BRITISH MUSEUM in the accompanying article, trying to debunk the idea of an 'Abominable Snowman', said that the track was made by a langur. Langurs frequented our bungalow, thundering as they ran across the bungalow's tin roof, leaving their small round tracks in the mud. I knew those tracks. I had chased langurs off my toys for years. No museum curator could convince me that the track in the picture was langur-made, even if changed by melting snow.

Unlike other Himalayan animals, langurs do not mind human observers. Often I watched them feed on leaves in the chestnut trees. Langurs are often seen around trees because their food is leaves. So what does a langur eat in the snow? Langurs are social animals, so I also knew that it would be out of character and equally out of habitat for one animal to be walking alone in the snow, and on a glacier, exposed to a snow leopard; forget that mister museum curator who studies dead animals—you need to know langur behaviour.

Knowing that langurs are social animals is important. A troop cares for its individuals. Adults take turns grooming each other, and they do so for hours. A baby spends time with females other than its mother. Little langurs snuggle against bigger ones on cold evenings and show respect for their elders, even play with them. A juvenile sometimes delightfully jumps on to an older langur's back, pretends to bite it, jumps off, runs around to the elder's head and, squealing, gives the animal a hug before scampering off.

Taking me out hunting, Grandpa had taught me to always think of what an animal needs and what other animals need that animal for. Get inside an animal's thinking, he said, think always of food because they need it every day. Think, too, of protection because they need that all the time. And at certain times, think of sex, and at those times animals often do not think clearly. If I, as a boy, knew these things, museum curators should too, especially at the famous British Museum.

In the library of Woodstock School the following Monday, I read everything I could to find about the Yeti. The next Saturday I asked Mother if I could walk down to the British Library in Mussoorie, above where the car traffic stopped and where we started the climb to our houses; the walk takes almost an hour. Not even Mother, a reading teacher, was sure her eleven-year-old boy would walk that far for books.

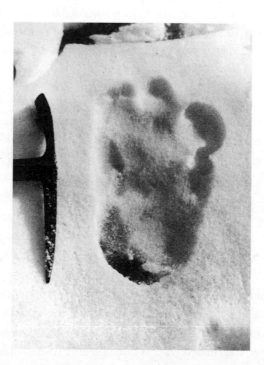

4.2   The Iconic 1951 Yeti Footprint Photographed by Eric Shipton

Source: Royal Geographical Society

I found nothing helpful at the library, but on the return, passing the bookstore on Mullingar Hill, sitting in the window was
Shipton's book, *The Mount Everest Reconnaissance Expedition 1951*.
I took it home with a promise to the shopkeeper that I'd bring the
money later. That hike took the rest of the afternoon as I read the
whole book, intrigued by Shipton's account:

> It was on one of the glaciers of the Menlung basin, at a height of
> 19,000 feet, that late one afternoon, we came across those curious
> footprints in the snow the report of which has caused a certain
> amount of public interest in this country. We did not follow them
> further than was convenient, a mile or so, for we were carrying
> heavy loads at the time, and besides we had reached a particularly
> interesting stage in the exploration of the basin. I have in the past
> found many sets of these curious footprints and have tried to follow
> them, but have always lost them on the moraine or rocks at the
> side of the glacier. These particular ones seemed to be very fresh,
> probably not more than twenty-four hours old.[1]

The following Saturday I went back to the bookstore. It was
then I became aware that the owner was a friend of the family's—
that was why on my first visit he had said that I could bring the
money later. Over following visits, he, too, developed an interest
in the Yeti and used announcements of new books to help my
search. We chatted in English, a bit in Hindi, for that was the
way we who lived in Mussoorie talked. The old man remembered
my father visiting his store as a boy; he was a high-caste Brahmin,
who, like many Brahmins, paid attention to lineage. As his
interest in my Yeti quest grew, he ordered titles I suspect he otherwise wouldn't have. 'Don't worry, Danny, I can always sell these
books by calling them mountain climbing,' he chuckled. 'White
men like to read of high exploits when they vacation in the hills.'
Indeed, after I returned each book, careful so it would not look
like it had been read, I would next see it in the store window under
'Latest Expedition'.

Looking back now I realize that in the old Brahmin I had a private librarian helping research my search. He died a few years after he first helped me and would have been very interested in my discoveries a quarter of a century later on that Barun ridge.

The books I brought home showed explorers were finding a lot. The 1952 Swiss Everest expedition (that narrowly missed making Everest's first ascent) found prints. That year the Belgian zoologist Bernard Heuvelmans published a paper theorizing that the Yeti was a relic population of an early hominoid, *Gigantopithecus*. According to him, a remnant of those prehistoric people had possibly retreated into Himalayan valleys escaping the Homo sapien populations in India and China as our human species started multiplying. Many accepted Heuvelmans's hypothesis. The Himalaya were a region of mystery, home of Shangri La, lost people in the mountains. His hypothesis built on something certainly true: the populations of China and India were growing. His hypothesis made a reasonable assumption—if Everest's high reaches had until then been untouched, the valleys of these mountains must have untrammelled valleys. Heuvelmans pointed to fossil evidence showing that *Gigantopithecus* truly existed fifty millennia ago. And with his proposal grounded in fossil evidence, the Yeti postulates no longer floated in the high altitudes of unknown origins. A pedigree was added to the footprints in the snow, grounded in the human evolutionary tree.

But where in these valleys was the Yeti hiding? Magazines were finding that Yeti stories attracted readers. Abbe Bordet found tracks at 12,350 feet, A.J.M. Smyth at 12,375 feet, and L.W. Davies at 12,000 feet. Charles Evans found tracks at 10,000 feet that his Sherpas claimed belonged to a Yeti—but interestingly, Evans started to critique his own discovery that it might be a bear and not a hominoid. Lining up the stories in my scrapbook, the evidence was a jumble, but all seemed to be presuming, in sober British fashion, that the Yeti existed as a real animal in some form.

The London *Daily Mail* had the year before, in November 1955, dispatched a Yeti hunting expedition. That seemed to be the first exclusive Yeti exploration. On it were a dozen Western scientists and 300 Nepali porters. Scouring eastern Nepal, they found six sets of Yeti tracks, most 8 inches long and 4 inches wide, as well as Yeti excrement—and in it was mouse hair, fur from an unknown animal, a feather, an insect claw, and plant products. That expedition concluded that the Yeti must be omnivorous, and of course anything that left scats was a real animal. Facts were confirming that some real animal lay behind the mystery, speculation was popping.

Some proposals connected to fables in other countries, some to legends of fantastic monsters. But all these sensational proposals concluded by grounding their postulates in the undeniable truth of animal footprints. In my eleven-year-old mind, these descriptions of the real with what was now a real possible paranormal brought new worries. Could a Yeti wander to our house on the jungle's edge? Were ape-men outside my room? If langurs and rhesus monkeys came, both primates like the Yeti, might wild men come too? Aunt Margaret's leopard shot from the back veranda proved that animals able to eat people were indeed just outside the door.

I pressed Mother to let me go into the jungle to begin my search. A month had passed since Dad was last with us having then taken a break from his medical work in the hot Indian plains. On that trip he had taken me into the jungle. So, in his absence, Mother gave permission to spend a night at a cave on the ridge below Childers with a shikari who she trusted and who sometimes accompanied Dad when he hunted ghorals.

It was late morning on a Saturday when Ram Lal and I set off. The monsoon poured, and we arrived at the cave soaked. The leeches were out. While Ram Lal started a fire to make tea, I slipped to the back of the cave, stripped off my shorts, and found a blood-swollen beast, thick as my thumb, inside my left thigh.

Another leech, still skinny, had hooked behind my right knee. I sprinkled a pinch of salt on each, and in my flashlight beam gleefully watched the bloaters writhe, salt wreaking havoc on their sensitive skin, then stamped on the creatures. Only after pulling on dry clothes did I see that my vengeful stamping had obliterated tracks that might have been in the cave. I played my flashlight anyway, looking for droppings.

Leeches fascinated Dad too. Each of those worms, *Hirudinea*, he told me, is both male and female, containing one pair of ovaries and nine sets of testes. But one animal cannot mate with itself. It takes two leeches to reproduce. Either embraces the other—it does not matter which—but the two line up with the front of one to the rear of the other, and from the rear of each leech sperm is injected into the other's mouth. Eggs are laid by the other who then takes on the role of the mother, wrapping them in a cocoon. From the moment of birth, these babies seek blood (if they are of the blood-seeking group, not invertebrate-eating leeches). Dad and I looked at leeches under his microscope. Each mouth has three jaws, not two. When leeches bite these jaws open the flesh with a Y cut. Then they bring the skin apart to create a hole and the tiny wound really bleeds; after that, with the leech mouth cupping over the flow, the giant gut, that is all a leech really is, sups.

When bloated, leeches release their bite, and as swollen grotesques they squirm to shade where the stolen blood thickens. Clear fluid is squeezed out through skin membranes. Thus digesting, soon a leech lies in a puddle of its making. No digestive enzymes exist in the belly, as with other animals, to do the work that follows. What is unique to this animal trapped between being both boy and girl, that has just concentrated its stolen blood into goo, is that a peptide in the gut starts to uncouple the amino acids in the blood, the process multiplying in the gut (it started as soon as the first drop of blood entered the body). As this fast-growing digestive system blossoms (enzymes, which most animals use, adequate to consume blood five times the volume of the animal would take up too much body space), the peptides break down the blood—the leech is distinctive, digestion without enzymes.

To humans, leeches seem to be little worms, living and squirming on the ground as they do. I once thought they must retreat into holes in the ground like worms. But leeches define everything in their unique way. They hide at the top of soil in wet cavities behind rocks and roots. Like tiny wisps that can fit almost anywhere, they remain hidden, secluded into moist depressions. They live for months on that bellyful of blood, enduring any heat short of fire, any cold short of freezing. Light, yes. The dark is also not a problem. What they cannot survive is dryness. Those in the blood-loving group wait, poised-to-bite, and when strengthened by a gut full of blood, they wait to attempt the role of one gender or another in sex.

One worry fixated my youthful mind; it was not the taking of blood, but a fear when I learnt how these animals thrive in wet cavities. Unsuspecting people or animals can drink from a stream and ingest a wisp of a leech swimming in that water. Uncomfortable in the acid of the stomach where it is first swallowed, the beast crawls from the stomach up the throat and crosses into the windpipe. Happy now in moist air, it attaches to the trachea. The air passage (a perfect, wet, warm, dark cave) closes with the increasingly chubby leech swelling with blood. Swollen to be as fat as a person's thumb inside one's throat, even a cow can be strangled by this animal that is many millions of times smaller. When Dad told me this, I never again questioned whether I should drink water straight from streams.

That evening by a fire at the mouth of the cave I squatted with Ram Lal talking of what animals might be out after the rains. Towards sunset the rain stopped, and under the billowing clouds golden rays unfolded across that Himalayan valley above the Aglar River. Leaving my friend to watch the rice and lentils we were cooking, I headed out; there was not much of a chance to find footprints after a rain, but I was always hopeful.

A fire-breasted flowerpecker dashed overhead, sparkling from the rain, iridescent in the slanting shafts of the ending day. On its blue-green breast a red dot blazed like a Christmas-tree light. The bird settled into a clump of mistletoe on an oak down the hill. When mistletoe berries are ripe, the flowerpecker will eat them and perpetuate the special cycle it shares with its host. Botanists have named the plant *Loranthus viscum* because after passing through the bird these berry seeds exit in a sticky form. To get those seeds off, the flowerpecker rubs its bottom against a tree bark. Seeds that fall to the ground do not grow for the parasitic mistletoe seed needs the bark of a broad-leafed tree to germinate, preferably a bark with moisture-laden moss. Rooted in that, the mistletoe grows to feed another fire-breasted flowerpecker, or perhaps the rarer scarlet-backed flowerpecker.

ᚦ

ONE DAY, WENDING MY WAY HOME from the bookseller where we spoke about jungle cabbages and kings, and whether pigs have wings, I stopped at an overlook. White cottages flock these Mussoorie ridges, perching their homes like birds on the branch-like Himalayan ridges. A generation before my grandparents, the British had this town built by the labour of coolies. Women and children were sent here for protection to escape the heat of India.

As I watched, a cloud ascended from the plains below, and from inside it came a bird with white stripes flashing the underside of its wings, an adult Himalayan griffon. Stolidly it coursed, turning, climbing, skimming a ridge, peering into the valley below. It seemed to be studying something in the valley, but was wise not to descend to this possible carcass, for in narrow valleys the griffon leaves the dead for smaller carrion such as crows. Perhaps a young griffon, not yet having developed its wing stripes, might descend, but it would soon learn that it cannot take off on a valley floor without a slope, especially when filled with pounds of just-eaten

meat. Stuck in the valley, the bird can be overtaken by a predator. To gain height to glide from, hopping uphill, hoping to avoid
being eaten, such a learning bird will climb the slope fifty feet or
more, then run downhill, searching for an updraught to carry it
out beyond the ridges.

The literature that day the bookseller and I had been discussing
circled around Rudyard Kipling (*The Jungle Book, Just So Stories,*
and *Kim*) and Edgar Rice Burroughs (*Tarzan of the Apes*). For most
children these books are fiction. For me, they were life-scripting,
and the bookseller was learning this. Mowgli and Tarzan both grew
up in the jungle as orphans, and for three generations the jungle
had been where my family lived, where we went for meat. What
for others was a world of writing was the life edge I ran to school.

In that jungle world, Mowgli also found animal friends. Am
I wild, Mowgli wondered, or am I human, bridging both and fitting
with neither. A prior night's dinner conversation revolved around
how a child might live in the jungle. The possibility was real
because Dad, when a three-month-old infant, just like Mowgli,
had been lifted from his cradle into the jaws of a she-wolf who had
lost her pups when villagers discovered her den.

My grandparents had come here from Kansas, ex-cowboys
turned medical doctors, and each year for six months the family
travelled through the jungle by an ox-cart, living in tents, dispens-
ing medical care. Jungle camps were not just a bridge to the wild,
but like Kim, they connected identities of white skin and brown.
A youthful Kim had showed me that language helped cross the
colour line, telling us white kids who wanted to blend in to get
behaviours as well as accent right if we wanted to move in this
land of India. Other dinner discussions were about how Mowgli
could communicate with animals that had no language. During
my talk with the bookseller, he kept reminding me that language
is more than words; it is first a respect for those with whom one is
talking, something I must give him since he was a Brahmin, and
a trait for India he was showing me that Kipling had. Kipling's

language sensitivity showed with words that people who bridged both languages would catch, as in Gunga Din's 'Hi Slippy, *hitherao!* Water get it!' ('Here' in Hindi is *idhar,* which bastardizes nicely through hither to 'get it'.)

Tarzan, like me, was a child of privileged parents. But people were not available who could teach him. He was raised by apes who had killed his parents. It was an intriguing idea: to be swinging from vines, learning from and by being part of our prehuman family, but knowing at the same time that one had to also be careful. Then in the jungle Tarzan discovers his parents' cabin and his father's knife. With a growing brain Tarzan struggles to grasp whether he is a human or a creature of the wild.

For an adolescent, the worlds I lived in were stacked one on top of the other—what is real and what is read, and what is true among that which is read? Like many adolescents, I tried to fit with the stories I read. Other American boys were playing cowboys and Indians; cowboys were what my grandparents had been, and Indians were the children with whom I played and whom my grandparents and my father cared for in their clinics. I was bridging worlds and finding my life propelling me towards a manhood with the wild. To quote Kipling:

> If you can dream—and not make dream your master;
> If you can think—and not make thoughts your aim;
> If you can meet with triumph and disaster;
> And treat those two imposters just the same ...
> Yours is the Earth and everything that's in it,
> And—which is more—you'll be a Man my son!

Watching the griffon soar, I pondered over my grandparents' early years. What was it like in the India of King George V? Most missionaries found their callings in colleges, hospitals, churches— a calling of changing others' allegiances. Doing so then was a lonely life away from their distant American homes. Living in the jungle must have been especially lonely, but it also gave us security and helped us get closer to the people and place where we lived. Many of those were years of the Great Depression, and money promised from America sometimes never came; the gun was useful then, and also my family ate food from villagers' fields brought in gratitude for the medical services rendered.

It took five months for a letter, carried by a runner, a rail, and a steamer, to make its way to the brother in Kansas or the sisters in Cincinnati, and months more for a reply to return to the jungle. Now in 1956 it still felt lonely, even though I could send a telegram to my other grandparents in Pennsylvania—where the switchboard in town would call them out on the farm and read over the party line with neighbours eagerly listening of news from India.

As the griffon soared on to other ridges I walked up the hill to our home. Entering adolescence, a white among browns, in a family that identified itself with a mission, questions of identity rode the storm currents of youth. The Brahmin had mentioned another Kipling line: 'If any question why we died, tell them, because our fathers lied.' It is easy to get stuck in valleys, hopping and trying to get lift, eating the carrion of others' lives, believing the lies of our forefathers when told sincerely. I would later find out that in America for most who were my age, the questions were not of racial identity but of sports and girls. But I was caught between a culture not mine and the jungle which felt like it was mine.

Halfway through the school year, as it happened, came the big assignment of fifth grade: we had to write a paper on any topic we wanted. At first I thought I'd do tigers—tigers were the ultimate quarry for Taylor boys—but I decided to write about Yetis. A paper that rang with this ending: 'Just one set of footprints need to be found. A thousand Yetis do not have to be found. All we need is one!' That paper began a lifelong research project.

THE YETI MADE ITS FIRST PUBLIC APPEARANCE in 1889 when an energetic British army physician, Major L.A. Waddell, returned from a hunt on the high Himalayan glaciers. He brought a trophy more sensational than the mountain sheep or bear he had gone seeking. Out of the Himalaya, Waddell brought sightings of footprints that ascended a glacier, then disappeared over a ridge—hominoid-like footprints. With those prints the Yeti first walked into the Western consciousness.

This end of the nineteenth century was also the beginning of the reign of science, the reign also of Queen Victoria when gentlemen explorers spanned the planet. Lands were being 'discovered'. Science and exploration moved together, uncovering mysteries throughout new territories. Charles Darwin's *On the Origin of Species* had revolutionized the understanding of how animals came to be, a theory that after its publication was more credible each year and became a defining vision for the Victorian man. So by the time of Waddell's report of the mysterious footprints, a global search had been underway to find the 'missing link', to explain, in the oversimplified perception of the time, how monkeys evolved into humans. It was appropriate that the missing link might hide in the most remote high snows.

New truths were being articulated about relationships with nature. 'New peoples' were being brought to the 'civilized' world. Fantastic postulates of the hypothesized were being proven. Indeed, science fiction was gaining respectability as a literary form. Anything seemed possible. Anything might not be fiction—even the Bible that at first seemed to contradict evolution, for other people had its inconsistencies newly understood with stone-written facts of archaeology. Informed people were learning to turn around earlier biases. For the open-minded, religion, too, was having misunderstandings of forefathers re-understood, joining the scientific age. In a world of discoveries, the missing link, while yet undiscovered, seemed ever more probable.

With the caution befitting a Victorian man of science, and careful not to overstep as an officer in the British army, Waddell did not categorize the footprints. Instead, he brought what he presented as another fact, the legend: '[O]f hairy wild men who are believed to live amongst the eternal snows, along with the mythical white lions, whose roar is reputed to be heard during storms. The belief in these creatures is universal....'[2] With those words in the colonial conquest of the unknown, the Yeti stepped forward.

In 1915, a letter had come to the Royal Geographical Society addressing these discussions. The letter was sent by the forestry officer J.R.P. Gent. It described the existence of

> ... another animal but cannot make out what it is, a big monkey or ape perhaps.... It is a beast of very high elevations and only comes down to Phalut in the cold weather. It is covered with longish hair, face also hairy, the ordinary yellowish-brown color of the Bengal monkey. Stands about 4 feet high and goes about on the ground chiefly. ...[3]

Unexplained fact after another was being newly understood, and so evidence was produced from an idea. Somewhere out there in the snows, the reports kept postulating, roamed the kin to humans. Some reports of the Yeti that were sent in, the gentlemen in their clubs discussed, were assuredly fabrication. Others must be mistaken identity, they told each other thoughtfully. Gentlemen formed learned societies to discuss these matters. The common characteristic of all these reports, though, was that they were only reports. Yet the reports kept mounting. Could all these reports from reliable, well-educated gentlemen be wrong? Were the stalwart villagers in valley after valley lying?

The name 'abominable snowman' was coined in 1921 when the Royal Geographical Society sent a well-equipped expedition to reconnoitre an ascent route for Mount Everest. Led by Lt Col C.K. Howard-Bury, the group approached Everest from Darjeeling, over the Cho La, and went into the spectacular Kama Valley. As they crested a pass they saw strange dark figures crossing another snowfield. Then later, they came across elongated, human-like tracks in the snow. Expedition porters claimed that the distant figures were *metoh kangmis* and the new-found tracks were made by these 'men of the snows'.

As this expedition was returning to England, in a public interview for the *Statesman* (the same newspaper where thirty-five years later I saw the footprint), a *Statesman* writer, Bill Newhouse, (who used the pen name Kim) saw a story in this discovery in addition to the Everest quest, and he translated *metoh* (which exactly means 'bear') first as 'filthy', then, not liking that, as 'abominable'. (*Kangmi* is indeed 'snowman'.) Moreover, Newhouse later confessed, 'The whole story seemed like such a joyous creation that I sent it to two or three newspapers.'[4]

4.3   A Yak Caravan Crossing the Same Pass in 1990 Where the 1921 Everest Reconnaissance Saw *metoh kangmi* or the Abominable Snowman

Source: Author

From 1921 until thirty-two years later when it was climbed, conquering Everest became a British fixation, and a discovery to accompany that, especially for the failed expeditions of the 1930s, was often the Yeti. As my father once quipped, 'If an expedition found their peak abominable, it was then proper to sight a Yeti.' Dad was a scientist. He drew my attention to one particular early piece of evidence, the only Yeti sighting made by a scientist. In 1925, an animal was seen by a member of the Royal Geographical Society. On the Zemu Glacier of Sikkim, N.A. Tombazi noticed his excited porters: 'Two to three hundred yards away and down the valley to the East of our camp. Unquestionably, the figure in outline was exactly like a human being, walking upright and stopping occasionally to uproot or pull at some dwarf rhododendron bushes. It showed up dark against the snow and, so far as I could make out wore no clothes.'[5]

Tombazi was applying the newly developing tool of photography to his scientific work. While he was attaching his telephoto lens (in the way that plagues so many 'almost-made' discoveries), the animal walked out of the camera's view and into the dwarf rhododendron. Disappointed, Tombazi rushed to the spot where he:

... examined the footprints which were clearly visible on the surface of the snow. They were similar in shape to those of a man, but only six to seven inches long and four inches wide at the broadest part of the foot. The marks of five distinct toes were perfectly clear.... The prints were undoubtedly of a biped, the order of the spoor having no characteristics whatever of any imaginable quadraped.[6]

Over the years these dimensions would be the most common with the huge one-foot-long prints found by Shipton being the other size.

In the 1930s, Wing Commander E.B. Beauman had found hominoid-looking tracks on a glacial snowfield at 14,000 feet. Ronald Kaulback came across tracks at 16,000 feet. On a glacier in the eastern Himalaya, Shipton came upon his first set of tracks.

In 1937, Yeti discoveries surged. H.W. Tilman encountered tracks on the Zemu Glacier just east of the closed land of Nepal (and the Barun Valley). On a snowfield at 19,000 feet, again on the Zemu Glacier, John Hunt (who would lead the successful ascent of Everest in 1953) discovered his Yeti tracks, prints similar to those seen by Tombazi.

So by the end of the 1937 climbing season, two expeditions had claimed to have seen the Yeti, seven had found its footprints, and many more descriptions had come in based on what local hunters and guides reported. Many of these reports mentioned Yeti hands and scalps in Tibetan monasteries. And Sherpas who climbed with westerners on the Tibetan side of Everest told especially of relics and sightings in Nepal's eastern valleys. But no westerner in the 1930s was permitted to enter Nepal.

These reports from the southern rim of Tibet confirmed stories emerging from Tibet's eastern side that described human-like animals in the bamboo. Were the Chinese 'wild men' the same as the mysterious soft skins, mostly white but with peculiar black spots, which had been showing up in China since 1869? Some animal wearing a white-and-black robe roamed the Tibetan Plateau. That animal's identity was revealed when Teddy and Kermit Roosevelt shot a specimen on one of their epic post-presidential expeditions: the giant panda. But people in China's mountains said that the panda was not the wild man. That was yet to be brought in—a man, they said, with skin, not an animal with fur. Through the 1920s, on both sides of Tibet, the search went on.

To explain the mystery, a discovery came from a back alley of Hong Kong. In 1934 a Dutch paleontologist, Ralph von Koenigswald, walked into a chemist's shop looking for fossils that were sold as 'dragon's teeth'. From the jars in which they were kept, he poured all sorts of fossils out on the counter. There, among the fossils, was a human, lower third molar, like no human tooth science had yet recorded—five or six times too large. It appeared to be the tooth of an extinct race of giants. For five years, von Koenigswald searched the China coast. In 1936, he found another, and in 1939, a third. Molars were hard facts; with these he could build out a skeletal description, and based on the teeth, *Gigantopithecus* was proposed. Giants 11–13 feet tall, they were fantastic omnivores who stood their ground in an ecosystem of sabre-toothed tigers and woolly mammoths. Von Koenigswald's teeth were the basis of Heuvelmans's Yeti hypothesis.

Then Frank Smythe made a discovery. Finding Yeti prints in a snowfield at 20,000 feet, Smythe pursued them, and although film was expensive, he photographed extensively. Back home from the high altitude, he analysed the prints using his field notes and comparative samples:

> On the level the foot marks averaged 12 to 13 inches in length and 6 inches in breadth, but downhill they averaged only 8 inches in length. The stride was some 1½ to 2 feet on the level, but considerably less uphill, and the foot marks were turned outward at about the same angle as a man's. There were well-defined imprints of five toes.[7]

At a high altitude, Smythe had concluded the tracks being so human-like that they must be made by a Yeti. But his analysis after recovering from fatigue and oxygen deprivation showed the tracks were by the Asiatic black bear *Selenarctos thibetanus*. He then presented a scientifically grounded Yeti option: quadrupedal bears could overprint hind paws into forepaw tracks and create bipedal-appearing tracks that vary in size—larger tracks when the hind foot fell back as the bear went uphill, and sometimes on the flat, and shorter tracks when the foot came forward on the overprint when the bear was going downhill. The larger prints fit Shipton and Ward's discovery, the smaller prints fit Tombazi's.

Notes

1. Eric Shipton, *The Mount Everest Reconnaissance Expedition 1951* (London: Hodden & Stoughton, 1952), p. 54.

2. L.A. Waddell, *Among the Himalayas* (London: Constable, 1900).
3. J.R.P. Gent, 'Letter to Royal Geographical Society', quoted in Bernard Heuvelmans, *On the Track of Unknown Animals* (New York: Hill & Wang, 1958), pp. 135–6.
4. Quoted in H.W. Tilman, *Mount Everest 1938* (Kathmandu: Pilgrim Publishing, n.d.), pp. 127–37.
5. Tombazi quoted in Heuvelmans, *On the Track of Unknown Animals*, p. 130.
6. Tombazi quoted in Heuvelmans, *On the Track of Unknown Animals*, p. 131.
7. Smythe quoted in Heuvelmans, *On the Track of Unknown Animals*, p. 134.

five
Yeti Expeditions

YAK EXCEPTIONS

Previous page:
5.1   A Yak Herder Leading the Author over a 17,000-Foot Pass behind Mount Everest

Source: Author

February 1950. I was five. Grandpa took me hunting one morning, not the usual out the backdoor but deeper into the jungle this time, and we jostled off across the Ganges past Hardwar in his World War II jeep. We spent the night in sleeping bags, awakened in the predawn dark, and I followed as he led me on foot quietly to a bank overlooking a dried-up riverbed. He stuck sprigs into a woollen hat, making his head look like the top of a bush.

I sat then, doing my part, looking through a peephole he had made in the bush, snuggled against the rough wool of my grandfather's jacket. 'Watch. Listen,' had been his instructions. 'Take the jungle inside as if you're drinking a glass of water; let the jungle flow through all your body.' That was the morning I learnt how a leopard stalks the edge of a riverbed. 'Feel this animal as part of you, don't be afraid,' he whispered. 'A leopard is beautiful; absorb how it walks a courageous and sly trail.'

Leopards are animals of the night. As the sun began to rise after the leopard was out of sight, down the riverbed followed three spotted deer, a wild boar, and finally we sat watching the ebb in India's rose-laced dawn. Moving to the riverbank, our legs hung over the sandy bank. Grandpa opened a thermos of hot sweet milk. As it washed into me, twigs still bobbing from his cap, Grandpa told me of his encounter with a man-eating tigress a few years before.

It seemed this certain tigress attempted a meal of a porcupine but was driven off by a swat of quills, of which three stuck, then festered, making the tigress so lame she started supplementing her diet with humans. Sportsmen who came trying to shoot her left birdshot in her rump and a slug in her shoulder, making her even further crippled and cagey.

Then one afternoon, two girls—one an excited bride everyone said was rather boisterous—were cutting fodder for their water buffalo. The bride, mocking her friend's fear of the man-eater, was climbing down the tree. Suddenly the tigress emerged from a bush, seized her, and disappeared into the jungle, her teeth piercing the thrashing, screaming girl's chest, holding her sideways in her mouth as a dog might carry a bone.

On hands and knees we tracked the tigress through tunnels of thorns and undergrowth. The trail was easy; we could see fragments of torn clothes, plenty of drag marks from the flailing girl, and lots of blood. Your father and uncle worked as though possessed, each day searching for fresher signs of the tigress, returning each night to watch at the spot where we had found the bride's hair and bones. I told them: 'Don't search where the tigress has been. Decide where it will likely come. Be there, wait.'

I thought that sooner or later the tigress would walk a particular trail, and there I tied my cot high in a tree and slept, trusting my feelings. On the fourth morning, I awoke sensing her impending arrival. What I heard were morning's sounds, but that sixth sense had made me sharp. No birds gave alarm calls. The sense is more than a gut feeling; it's a developed sense I've had a number of times. When it comes I am as certain as if from a smell or a sound. In that cold light that grows out of the jungle night, before the sun pokes over the horizon, the day was stepping alive, like the moment this morning when those wild boar crossed this riverbed.

At that time of change the tigress limped down the trail, the first sun rays strong on her golden sides, shadows streaking her black stripes. One shot, and I killed her—it was with my .405 magnum, Teddy Roosevelt's favourite calibre. That tigress died for that bride.

Such experiences were parts of my childhood training. 'Watch. Listen. Absorb.' We went to the jungle for meat, but even more to be with the jungle, often just drinking deeply from being with it. The jungle is a complex world where understanding comes by living with it.

Six-and-a-half years after that jungle morning my Yeti quest began. Several weeks after seeing the picture in the newspaper, it occurred to me to do what we did when hunting other animals. Ask the people. Taking my Shipton book, I walked to the coolie stand below Sisters Bazaar where a dozen hopefuls sat. They were from villages throughout our hills, people who travelled Himalayan trails all year. Like travellers through the ages who move without maps and guidebooks, descriptions of the road are shared at resting places, telling each other of turns, washouts, changes, and dangers. Mental maps become updated.

I showed Shipton's photograph to a coolie who seemed to be from back in the mountains, jute cloak stained with black streaks from carrying charcoal in 150-pound loads.

'Yes, I know what made that footprint. That is a man's. Don't be confused by round toe marks or toes together. In snow a man's toes look round, not long as in dirt or mud.'

'The print cannot be a man's,' I said. 'This picture was taken on a snowfield no person had travelled across for many months.'

'You are still a boy. You do not know that in these mountains, even high up, people always travel. Never think no one has passed. Maybe the maker was a holy man on pilgrimage—pilgrims are always roaming our mountains. Sometimes people also travel in secret, carrying things they do not want others to find. Maybe this person worked for a government and was going north into Tibet? In these mountains unknown tracks are found.'

'But coolie*ji*, look at how big the footprint is. That axe-head beside it shows the footprint is more than 12 inches long.'

'Ah, I did not know the footprint to be so big. So that means an unusual man. Unusual men walk in unusual places.'

'Coolieji, have you ever seen a Yeti's footprint? Is this the track of a Yeti?'

'What is a Yeti?' he asked.

'The Yeti is a wild man that lives in high mountains,' I replied. 'It has long hair and a pointed head.'

'Such men don't live in our mountains.' While the charcoal porter and I were talking, others had crowded in and began examining the picture. Nods followed.

That was the first day I had actually asked any villagers about the Yeti. And all had said 'no'. Why did they not admit knowing about it? Was their lack of admission similar to what they said when asked about evil spirits, beings I knew villagers also believed in but did not discuss. That was the day I first learnt that my Himalaya were not Yeti-land. Mussoorie is in the India Himalaya. The land of Yetis is 600 miles away, over a hundred Himalayan passes in eastern Nepal.

That conversation with the charcoal carriers at Sisters Bazaar was twenty-seven years before I would find my footprints in the Barun snows, but I had begun exploring valleys and ideas. Grandpa had taught me the patience of the hunt and that I must learn the hunter's skills. The greatest hunter of our jungles was Jim Corbett, whose books my family studied as textbooks and not just thrilling jungle stories; his home range was adjacent to where our family roamed. It was he Grandpa was quoting when speaking of 'the sixth sense'. Here's an excerpt from his *Man-Eaters of Kumaon*:

I have made mention elsewhere of the sense that warns us of impending danger, and will not labour the subject further beyond stating that this sense is a very real one and that I do not know, and therefore cannot explain, what brings it into operation. On this occasion I had neither heard nor seen the tigress, nor had

I received any indication from bird or beast of her presence and yet I knew, without any shadow of doubt, that she was lying up for me among the rocks. I had been out for many hours that day and had covered many miles of jungle with unflagging caution, but without one moment's unease, and then, on cresting the ridge, and coming in sight of the rock, I knew they held danger for me, and this knowledge was confirmed a few minutes later by the kakar's warning call to the jungle folk, and by my find the man-eater's pugmarks superimposed on my footprints.[1]

Or, in another reference of sixth sense from the same book:

As I stepped clear of the giant slate, I looked behind me and over my right shoulder and—looked straight into the tiger's face.

I would like you to have a clear picture of the situation.

5.2   Grandpa in 1923 with a Leopard He Shot as It Had Leapt at Him

Source: Taylor family archives

The rifle was in my right hand held diagonally across my chest, with the safety-catch off, and in order to get it to bear on the tigress the muzzle would have to be swung round three-quarters of a circle. The movement of swinging round the rifle, with one hand, was begun.... Only a little further now for the muzzle to go, and the tigress—who had not once taken her eyes off mine—was still looking up at me, with the pleased expression still on her face.

How long it took the rifle to make the three-quarter circle, I am not in a position to say. To me, looking into the tigress's eyes and unable therefore to follow the movement of the barrel, it appeared that my arm was paralysed, and that the swing would never be completed. I heard the report.... For a perceptible fraction of time the tigress remained perfectly still, and then, very slowly, her head sank on to her outstretched paws.[2]

☙

MONTHS AFTER I DISCOVERED THE YETI IN A NEWSPAPER, Dad received a grant from the Upjohn Pharmaceutical Company to search for a plant, *nusha bhoota*, meaning 'the hair of the ghost'. In the Kulu Valley, when the flower is in bloom during the monsoon, shepherds pass carefully through high meadows, for at these times the *nusha bhoota's* blossoms emit vapours that allegedly anaesthetize passing travellers who then collapse. Upjohn thought the plant could be used to make a new drug. Uncle Gordon sceptically said one night at our dinner table that maybe the travellers fall asleep because after carrying heavy packs on a sunny day, lying for a while in a warm pasture is pleasant; then when monsoon clouds blow in and the rain hits their faces, the sleepers awake.

I was pressing to join this expedition because earlier Uncle Gordon had said to Dad with a very sincere voice, 'Carl, keep your eyes open. Those high pastures in the monsoon could be Yeti habitat. You might have more luck finding the Yeti than *nusha bhoota*'; if jungle-savvy Gordon thought the Kulu Valley was Yeti-land, I wanted to go.

Arriving at those high meadows in a different part of the Himalaya, what we really found was rain. Wind-driven rain slammed in sideways. The higher we climbed, the thinner the rain turned, but there always was rain. No matter how high we climbed—10,000 feet, then 12,000—we could not climb out of the rain. In evenings we stood in clammy clothes trying to dry ourselves off by smoky fires, or lay in our tents in soggy sleeping bags. Moisture from our breaths struck the cold fabric in those early alpine tents, then condensed, and caused it to rain inside the tent.

Each morning before the scientists left we shared tasteless oatmeal and sweet tea by the fire. After they left, I would clean the cages and feed the white mice and guinea pigs we used to test the anaesthetic qualities of each plant. Afterwards, there was not much to do. One morning, sitting under an umbrella by the cookfire, smoke swirled in, mimicking the clouds over the ridge. The smoke swirls then re-spun, heading on over the ridges.

Life that unrolls flat on the plains gets reshaped in the mountains where earth and clouds fold over on top of each other. The sky sometimes lies below the earth, and the ground lands above the sky. In that world we stand in the sky with clouds below and clouds above, but we stand on land. Distance opens with views across mountain ranges, then the clouds coming in bring vision very immediate with the land on which one stands feeling like an island in the sky. Ideas come in like the clouds, embrace our understanding very closely, and then dance on. Ideas dancing with sky; the land dancing with clouds. In that miasma something happens to the mind, especially at altitudes. World and Air dance as partners, dancing closely as Earth and Sky. Mountains lose their rockiness. It is to seek such oneness joined to otherness that sages go to the mountains. And young boys, too.

A Tibetan shepherd stopped by the camp that morning carrying two heavy sacks, and on arriving let them slide off his back, whistling. Then, with a fast twirl of an oiled canvas, he covered them from the rain.

'I carry supplies into the Lahaul Valley,' he replies to my question, but when I ask what is in the bags, he falls quiet. Our cook hands him a mug of hot sweet tea.

I point to my mice and guinea pigs. 'My job is to care for those animals. Are there dangerous animals on this mountain that might attack mine?'

'Yes. Leopards could, bears also. After the monsoon bears dig for wild mice. Bears look into the ground for little animals. They will smell the ones you have.'

'Any other animals?'

He wasn't interested in possible predators for my mice, but the cook refilled his mug; that reminded him that our camp was more interesting than the trail. 'Martens and weasels, maybe you should be careful of even an eagle. It could see through the wire of their cages. You should cover their cages with cloth.'

'Are there manlike animals, wild jungle men?'

'I do not understand. What do you mean by "jungle men"? What would men-animals eat here on these empty ridges?'

'Maybe they feed on grass in meadows such as this. Maybe they eat animals like my mice. Do you know of the Yeti?'

'I have never heard of such men. We do not have any around here. Do not fear wild men. But watch for bears.'

And so in a second region of the Himalaya, as with every other person with whom I'd inquired of on that expedition, no one knew of the Yeti. Why were mountaineers and explorers finding Yetis? I thought more on the food question. Could the snowman be a grass man? For half of the year the high slopes are covered by snow; at those times what would a Yeti's food be?

When Dad returned to the camp that evening, he wondered: 'Today I saw a valley of rose-pink primulas. Yesterday I was in the valley east of it; the primulas were dark purple. The two colours must mean that they are two different species, for the purple primulas have that powder, farina, on their blossoms, and the rose-pink lack the white powder. Why do two species exist as monocultures in adjacent valleys but are not intermixing? Both valleys are at 13,000 feet. Both appear identical in soil, face north, and are similarly moist. Why does one have pink flowers, the other dark purple, and each valley have none of the other?'

Possible explanations were well along as we ate our curried carrots and cabbage. Each evening the last to join dinner was Vaid, one of our botanists, who had to interrupt his day's pressing of plants. 'Carl, the rose-pink is *Primula rosea*, and the dark purple is *Primula macrophylla*. I have been seeing these too. One explanation for why you found monocultures is a theory advanced by Kingdon-Ward in the far eastern Himalaya. After walking one valley and finding flowers of one colour and, like you, walking the next day in another valley and finding a second colour, he dismissed soil type and slope exposure as an explanation, noticing two features. One is the colour difference you noticed; the other is that the flowers are so abundant in the first place, for pristine meadows should show relatively fewer flowers as untouched meadows are so thick with grass that it chokes out flowers. When a valley is lush with flowers, it speaks of being grazed by domestic animals. And I am sure you've noticed that all the meadows we've been searching have their grasses chewed down.

'So, when you see a valley with a carpet of flowers you know you are not the first to visit this season. Because flower stems grow more quickly than grass, meadows with flowers were prepared for your viewing pleasure by shepherds weeks earlier. Recognizing then that grazing brought the flowers, Kingdon-Ward linked this to the colour difference. One colour dominates when grass is grazed earlier, another when grazed later. It remains a hypothesis, but it is a wonderful option.'

In my ardour in which I nearly revered the wild, I had never thought people could make Nature more beautiful than when it pristinely grew. This suggested that domestic goats, which I had thought to be rapacious eaters, were in fact causing splendid wildflowers to grow in profusion.

New ideas were coming to me. Growing up with scientists and villagers, talking English with one, Hindi with another, questions flooded my mind. Why did my native language, English, which my culture claimed to be more civilized, not use honorifics, but the Hindi of the people who, as missionaries, we were supposedly teaching not only had elaborate honorifics but also embellished salutations. Some months earlier Dad had punished me for using the familiar *thum* when I addressed an older woman. Impertinently I had then asked, after one of my few life whippings, 'What honorific should I use when telling a snake to get off the trail?'

The next day after learning that people made the wild more beautiful, Vaid and I were talking. 'Vaidji, some plants we collect are valuable. Are they hard to find?'

'People here have experimented for centuries; experiments that worked became plants that are hunted. Do you know a plant that looks like a coffee plant? You've seen it around Mussoorie; it is called *Rauvolfia serpentina*. Dark-glossy, green leaves with red flowers found in sandy gravel where there is forest cover. Its power is in its roots, long tubers which the local people grind into paste.'

'What do they use it for?'

'Madness, but also for a snakebite, birthing pain, and grief after a death. Scientific tests have shown a positive effect on hypertension. It works so well that drug companies are buying wild *Rauvolfia*. In many valleys, in slack parts of the agricultural season, people are digging up the bushes. Losing the bushes across many valleys may, of course, exterminate the wild *Rauvolfia*.'

'Why don't the companies just grow the plants in fields?'

'Growing one bush takes years. Then to harvest it must be killed. It is cheaper to pay villagers to harvest jungle *Rauvolfia*.'

'Well, villagers are making money. They are poor.'

'How much really? Maybe on a good day they triple their income over a day in the fields. When the bushes are gone, the villagers, who developed the knowledge over generations, will not be better off; then income is gone as well as the plant. That is why we are here; your father has convinced a drug company that maybe the *nusha bhoota* could be grown as a medicine.'

Two weeks later, our expedition left with a plant whose vapours killed one of my guinea pigs and three white mice, a plant the Upjohn Company could never get to grow in Michigan green-houses. Back in school, the biology we studied held much less interest than Saturday and holiday trips that took me into the field, and the readings I spent time with were Jim Corbett, teaching me the languages of the wild, not those of memorizing names where an interesting animal was described with a Latin name.

> Tigers do not know that human beings have no sense of smell, and when a tiger becomes a man-eater it treats human beings exactly as it treats wild animals, that is, it approaches its intended victims up-wind, or lies up in wait for them down-wind.
>
> The significance of this will be apparent when it is realized that while the sportsman is trying to spot the tiger, the tiger is in all probability trying to stalk the sportsman, or is lying up in wait for him. This contest, owing to the tiger's height, colouring, and ability to move without making a sound, would be very unequal were it not for the wind-factor operating in favour of the sportsman....
>
> For example, assuming that the sportsman has to proceed, owning to the nature of the ground, in the direction from which the wind is blowing, the danger would lie behind him, where he would be least able to deal with it, but by frequently tacking across the wind he could keep the danger alternately to the right and left of him. In print this scheme may not appear very attractive, but in practice it works; and, short of walking backwards, I do not know of a better way or safer method of going up-wind through dense cover in which a hungry man-eater is lurking.[3]

ⱺⱺ

LEAVING INDIA, I TRACKED THE YETI from the United States through my teenage years. Its trail was in column-completing stories in the middle of newspapers, reports that had changed from happenstance discoveries by mountaineers to major, moneyed, and scientific Yeti expeditions.

In 1957, 1958, and 1959, the Texas millionaire Tom Slick penetrated the long-closed valleys of east Nepal, particularly south of Everest as well as the Arun Valley. Slick's first expedition found three sets of Yeti footprints, hair, and excrement. The second used bluetick bloodhounds, trained in Arizona on mountain lions and bears. These canines required jackets to protect them from the cold, and during the monsoon needed lanolin to be rubbed daily into their paws to prevent cracking as well as American medicines that were fed daily to them to ward off exotic diseases. During four months in the field, this second expedition found another set of Yeti tracks and visited four remote monasteries to study two alleged Yeti scalps and a mummified Yeti hand. Slick announced he was on the verge of a major discovery.

He returned on a third expedition. Encompassing nine months, trekking a claimed 1,000 miles, the members trapped two black bears and a leopard, found numerous Yeti tracks, collected droppings, and got repeated confirmation of the animal from villagers. On returning to the USA, Slick suggested that maybe there existed three Yetis: one supernatural and two real. Of the two real, one fit the Tombazi report, and the other with large footprints fit with the prints seen by Shipton and Ward. Wealthy, and not accountable to a donor or museum, Slick would not release his evidence, saying he had to work through the relationships among the three. But before those findings came, he died in an aeroplane crash. Here is Peter Byrne's account of one of their discoveries:

> Ajeeba was ahead, breaking trail and we were deep in a grove of twisted rhododendrons when he suddenly stopped and pointed at the snow directly ahead. Tracks! Tracks that looked like the tracks of a man. Tracks in a place where there were no men but ourselves.

Running diagonally across our route was a set of deep footprints in the snow ... I came up quickly with Ang Dawa and Gyalzen and at a glance saw they were not the tracks of a human being, but footprints exactly similar to those that I had now come to regard as belonging to the yeti. ...

The tracks were still fresh. They had been made that morning, probably only an hour before our arrival and though I was keen to start following them at once, I decided to await the arrival of the Walung village men and see what their reaction would be.

One by one they came in and dropped their loads. I called them up to see the tracks. As each one approached, I pointed to the footprints and said, 'Look a *thom* has been here.' *Thom* is the Sherpa name for a bear. Together they examined the tracks and then without exception declared, 'No sahib, these are not the tracks of a *thom*. A yeti made these tracks. See, there are no claw marks and if a *thom* had made these, there would have to be claw marks.'[4]

Yeti hunting was attracting attention even from the government of Nepal which was increasingly having ethical, economic, and national identity issues as expeditions scoured the country's valleys for this wild man. So Nepal implemented Yeti-hunting regulations. The American Embassy, also taking Yeti hunting seriously, officially publicized these regulations.

1. Royalty of Rs. 5000/- Indian currency (rupees) will have to be paid to His Majesty's Government of Nepal for a permit to carry out an expedition in search of 'Yeti'.
2. In case 'Yeti' is traced it can be photographed or caught alive but it must not be killed or shot at except in an emergency arising out of self-defense. All photographs taken of the animal, the creature itself if captured alive or dead, must be surrendered to the Government of Nepal at the earliest time.
3. News and reports throwing light on the actual existence of the creature must be submitted to the Government of Nepal as soon as they are available and must not in any way be given out to the Press or Reporters for publicity without the permission of the Government of Nepal.[5]

The most noteworthy thing about these regulations was that the government of Nepal did not really have an idea whether the Yeti existed. Might it even be one of their people, a lost tribe? The implication, though, is clear: governments of both Nepal and the United States recognized that the Yeti might be real. In a sense, with such government regulations, the Yeti had acquired citizenship. The expeditions were bringing back evidence, but also big questions.

Then the most-publicized-ever Yeti expedition departed. In 1961, *World Book Encyclopedia* funded Sir Edmund Hillary and a multidisciplinary team to spend a year doing high-altitude research. The *National Geographic* sent Barry Bishop. An endeavour sponsored by *World Book Encyclopedia*, involving a staff member of the *National Geographic* and led by Sir Edmund Hillary said to me that my search was not a 'fringe' endeavour. Although the expedition's quest went beyond the Yeti, and members specifically downplayed that aspect, it was the possibility of finding the Yeti that fired news stories (and seemed to have motivated *World Book Encyclopedia* to pay the bill). *World Book Encyclopedia* had never sent any other expedition to the distant valleys of the world—and a further sign of credibility was shown in their assignment to spend a year in the field. Specifically, seriousness about discovering the Yeti was shown in that Michael Ward—Shipton's former partner—was a member. As Hillary himself was to describe:

> The search for the yeti created tremendous interest in the United States. While everyone agreed that little was known about the supposed creature, there were some rumors that it was half-animal and half-human. One newspaper in Chicago humorously suggested that if the yeti were brought back to that city a decision would have to be made whether to put it in the Lincoln Park Zoo or check it into the Hilton.[6]

The evidence the expedition brought home, though, did not support such conclusions. The revered scalp from Khumjung

Monastery was borrowed, and then sent for analysis to the Field Museum of Natural History in Chicago. Evaluation concluded that the scalp was a piece of serow (goat antelope) skin stretched into a pointed cap. While they did not bring any physical evidence, their cultural investigation was perhaps more damning, stating that the misunderstanding of the Yeti came from Westerners not properly perceiving how Sherpas view the Yeti. To the expedition's dismissals of the Yeti, sceptics responded that the team had approached that inquiry inappropriately; no anthropologist was on the team and none of the researchers spoke the Sherpa language. But Hillary directly addressed the cultural factor in his book:

> To a Sherpa, the ability of a yeti to make himself invisible at will is just as important a part of his [the yeti's] description as its probable shape and size.... Pleasant though we felt it would be to believe in the existence ... the members of my expedition— doctors, scientists, zoologists, and mountaineers alike—could not in all conscience view it as more than a fascinating fairy tale ... molded by superstition, and enthusiastically nurtured by Western expeditions.[7]

I found Hillary's conclusion interesting. Like so many mountaineers, conviction in the presence of the Yeti had been his starting thought. In 1953, after ascending Everest with Tenzing Norgay, he had mentioned to the media 'mysterious large footprints' found in the snow. Tenzing had added, in the press and later in conversations with Bob Fleming, how his father had seen a Yeti twice while out herding yaks. Now Hillary was being cautious, as was Tenzing in his republished, slightly edited new edition of his autobiography, *Tiger of the Snows*.

But this caution from the conqueror of Everest did not persuade a still hopeful world. Primatologist William C. Osman Hill, from the Zoological Society of London, then issuing an authoritative multivolume series on the comparative anatomy and taxonomy of primates, termed Hillary's denial 'rather hasty', and offered the alternative that the Yeti was a 'plantigrade mammal capable of bipedal progression', refocusing the hunt out of the high snows. According to him, 'Searchers for the snowman have been looking in the wrong places.... [The Yeti's] permanent home is undoubtedly the dense rhododendron thickets of the lower parts of the valley and it is here that future search should be directed.'[8]

၆ဝ

IN 1961, I ALSO RETURNED TO THE HIMALAYA, finally to Nepal, the home of the Yeti. Mother was making a movie on Nepali women; my job was recording the soundtrack. Our location in central Nepal was the first place I'd been in the Himalaya where I could not speak the language. Finally, I found a schoolteacher who spoke Hindi, and he told me how he learnt about the Yetis. He had picked up a book on Nepal written for tourists and while reading it, he had wondered what that author meant because there was no such animal where he lived in Nepal. He explained to me, 'Our word for Yeti is *bun manchi* [jungle man].'

The following week, a group of Hindi-speaking traders entered a tea stall where I was present. Their home was above the Marsyangdi River on the north side of Annapurna. They had been working in India as watchmen and were now headed home. I asked if they had bun manchi in their villages. And they said yes; bun manchi were in the jungles below their villages, and sometimes came even into their villages, but usually they stayed in the jungles.

I was making progress. Knowing the language was important, and knowing things about the culture, too, was essential. I recalled that Hillary's team was using translators. My Nepali language skills did not exist then, but I was getting information using Hindi, and I asked the Marsyandi men what their jungle men looked like. Their description was perfect. Adult bun manchi are shorter than people, with much hair. They never travel in the day. They like coming into cornfields. When they come, it is often possible to hear them screaming; they apparently have a high-pitched, long scream like that of an eagle. But these traders had never seen one of these. They were adamant that it was best to leave the bun manchi alone, for if you bother them they steal your children.

෧෪

NEPAL WAS STILL A SOMEWHAT CLOSED COUNTRY THEN. The valleys of Yeti discoveries lay along the border with China. Foreigners could go to Everest, but not the adjacent valleys because of fears from Cold War game-playing. American CIA operatives were active along this border, so were the Indian CID. They were supporting Tibetan guerrilla fighters who were working for the Dalai Lama fighting an insurgency. Because of this, the high jungles where Hill suggested the search should concentrate would stay locked off for years. And being locked off, absent new discoveries, Sir Edmund's conclusion dampened the continuing efforts to answer the puzzle.

But in 1969, entry into even the restricted valleys became possible for me. I returned to Nepal as a member of the US foreign aid programme. My job was to criss-cross the country in the family planning programme. I had a permit that allowed me to visit all seventy-five districts of the country. I even had a helicopter when on official business, so I could get in and out of remote places. Any valley where there were people had potential family planning need–giving me permission to enter them.

Again, I found language an important tool in Yeti hunting, and living in Nepal, I was learning Nepali. Usually when I mentioned this mystery to people, they would respond, 'Yes, the Yeti is here.' When I asked for a list of animals without mentioning the Yeti, the villagers never mentioned the one I wanted. Most lists, though, mentioned the bun manchi. Some 'thing' was eating real crops in their real fields. This come-in-the-night animal especially loved cornfields.

Easily accessible from Kathmandu was the Gosain Kund Ridge; it was a five-day trek one-way, but a twenty-minute helicopter ride. Short flights were a personal extravagance my salary allowed. When periodic Nepali holidays piggybacked with a Western weekend I would be dropped in one valley, with four days and five nights to cross to an adjoining valley for an early pickup, a quick shower, and being at my desk a couple hours late, fresh from a high-altitude sunrise, figuring out how to motivate men to adopt condoms or vasectomies.

On one trip I bivouacked at 16,000 feet. As a tent added weight, my practice was to use a nylon bivvy bag into which each night I stuffed a sleeping bag, boots, a water bottle, a flashlight, and myself. That night's bivouac was under a dramatically overhanging rock, my pack outside.

Snow started during the night, in thick heavy flakes. It seldom snows in early October in that part of the Himalaya. Maybe an inch, I expected, would dust the valley the next morning, ideal for tracking animals. But snow fell throughout that night. At dawn, from the overhang of my rock snow loomed from the ground to the rock's ceiling. Punching out into the snow, I found myself in a blizzard that was ending. My watch said ten o'clock. Ploughing to where I thought my pack was, I groped back and forth, probing. Could a Yeti have carried it? Hundreds of dollars worth of Nikon cameras and lenses were in that pack.

The helicopter was due shortly after dawn the next day. I had to hasten on to that site, for it would never find me on this side of the pass. Stuffing my sleeping bag, bivvy bag, empty water bottle, and stove into the pillowcase, and tying it with the parachute cord from my pocket, I started off. I would come back after the snows melted to collect the cameras. Two strides up, I stumbled into my pack. After a can of sardines in mustard sauce, I pressed on.

At three o'clock, I topped the pass. Descending into the other drainage, spying a high, flat-topped boulder, I spread my bivvy bag in the sun to dry and got the stove going. Sitting on the rock, swirling hot tea, I knew animals had their homes under the snow. For them, snow gave a blanket that kept predators away. Thermometers in tunnels they burrowed out register 60°F when temperatures above were below freezing. This subnivean world is filled with life: insects, especially spiders, and voles chasing the insects. Mice eat the vegetation. Pikas are the big beasts in this system, sleeping a lot under the earth, but when they need to eat in their snowy world they look for caches of stored food, covered from the eyes of searching hawks.

Might William C. Osman Hill be wrong in asserting that Yetis must live in the jungle because food exists only there? He worked in hot, steamy Ceylon. Do Yetis know about these snow denizens? Polar bears live in snow and ice. Might Yetis have found ways to harvest the reclusive larders of pika food, maybe eating the pikas also? And in some Himalayan valleys there are also marmots.

Under the earth is another world of even smaller dimensions. Worms twist and burrow. Beetles chew. Ants scurry. In their daily quests, none of them know about the animals above. The reality of each of these is the world before their noses. Inside each of these—beetles, worms, ants—are smaller eco-worlds, and life grows smaller as bacteria burrow through tissue. Microbes thrive in systems seemingly so small as to be without dimension. Even the more minute DNA holds characteristics. We deny consciousness to these life forms, but we could also be simply not yet informed.

Smaller than us, it is clear that we live in worlds stacked on worlds. Why not then do these worlds extend up higher? Or, to engage this from the other direction, what is reality beyond us? To think that there is none is conceit. People create our burrows and airways tunnelling through the cumuli of our lives. We think this to be the whole world, not giving consciousness to others where, in some cases, we know that consciousness exists. Consider the elephant and its intergenerational mindfulness. It is a mammal, not all that dissimilar from us—except that it keeps its nose in its own business and not in reshaping the planet.

On top of our world could be other worlds. For if smaller worlds exist, why cannot there be larger; that is, if we admit our ignorance. Paul Tillich talks about the 'ground of all being', awareness of the great that is so much more than ourselves, being sensed by our existential dynamics inside? As we believe ourselves the most enlightened of beings, might not our human limitation be limited like that of the parasite in the bowels of a pika? Might we be being affected by a larger being but can have no appreciation of that organism. Could there not so be a ground of all being? One name for it is infinity—we know such exists, but we cannot define it. Our universe expands to what we know not, but we know it is expanding. That we know not does not mean that there is not something out there. What might the mysteries of life animate into as they leave the present?

Another parameter for this, of course, is wildness—the world beyond our domestication. I have been searching for a man in the wild now, for a decade-and-a-half. It will be another decade-and-a-half before I discover the Barun footprints, nests, and two bears. But it is already clear that the Yeti is more than a wild man, even if it is not that. The Yeti is also a calling. How much of it is the wildness inside people questing for a connection with the wild that is calling from where our species once came? This would answer Hillary's claim in a more positive way that the Yeti is a Western creation. The quest that is driving our seeking could be to understand that great beyond.

୧ତ

THE FOLLOWING MORNING, I WAS IN A MEADOW of *Primula denticulata*—small, purple-mauve flowers widespread at that altitude. While milk was heating on the stove for my cereal, the sound of French-made helicopter blades suddenly approached me. An hour later, I sat at my desk reading the weekend cable traffic from Washington. A new book had come with the diplomatic pouch. That evening in 1970 at my Kathmandu home I started reading. Eric Shipton had again joined the Yeti discussion, adding unreleased details of his 1951 discovery with Michael Ward:

At 4 o'clock we were astonished to see a line of tracks converging on the direction of our advance, and evidently coming from or going toward the col at the head of the glacier.... When we reached the tracks we saw that they were fresh, certainly not more than a few hours old. ...

Sen Tensing was the only one of us who had no doubts as to the origins of the mysterious tracks. With complete confidence he pronounced that they had been made by Yeti ('Abominable Snowmen'). He told us that two years before he had seen one of these creatures at a distance of twenty-five yards. He described it as being the height of an average man, tail-less, with a tall, pointed head and covered with reddish-brown hair except on its face which was bare....

We followed these down the glacier. Gradually, as we descended the depth of the snow diminished, until there was barely an inch covering the glacier ice. Hitherto the individual footprints had been rather shapeless, but here we found many specimens so sharply defined that they could hardly have been clearer had they been made in wax. We could tell, both by comparing one print against another and by their clean-cut outline, that there had been no distortion by melting; and this again provided ample evidence that they had been very recently made....

We found several places where the creatures had leapt over small crevasses and where we could see clearly that they had dug in their toes into the snow to prevent their feet from slipping back....

Night was falling when we reached a moraine at the side of the glacier, where we camped. It was a clear, still night, and when we had settled in our sleeping bags the silence was broken only by the occasional creaking caused by the movement of the glacier. I could not altogether suppress an eerie feeling at the thought that somewhere in that moonlit silence the strange creatures that had preceded us down the glacier were lurking. I was not surprised to find that Sen Tensing, lying beside me, was pondering similar thoughts.

'You know, Sahib,' he said, 'the *Yeti* will be very frightened tonight.'

'Why?' I asked.

'Well, no one has ever been here before; we will certainly have scared them.'[9]

Putting the book aside, I stepped outside. It was late evening as I walked the Kathmandu streets with homes and stores shut. The only open place I passed was a temple with lights that

5.3 Lamps at a Kathmandu Altar

Source: Author

illuminated the idols and lamps inside—the Yeti is an idol, I realized, a spirit that gets incarnated into a material being. It has transcended its once supposed home in Buddhist lands. I walked the streets of Kathmandu, the city where the Yeti now lived with greater vitality than anywhere else, even the high Himalayan snows. It lived on T-shirts. It had become a brand of soap. And, of course, it was both a beer and a whisky. It was interesting that an animal of the most remote mountains found its life most vibrant now in a city—in this city it had become an idol that makes money.

In Shipton's new report, the facts are the same between his first book and this written in 1969. But across years when so many Yeti expeditions have taken to the search because of his discovery, Shipton has now discovered more, symbols like paraphernalia in the stores. Most interesting is that while the questers for the Yeti never knew what it looked like, and all they had was footprints, among marketers there is no disagreement as to its physical appearance. What is equally interesting is the total absence of any reproduction of its footprints in the markets.

But for us who still search, there is that iconic track to be explained. And around it grows the hope, transmogrified in this new book in the imagination of men lying in their sleeping bags on glaciers who remember their earlier discoveries: our wild ancestors left that form imprinted in the snow. As I walked through the streets of Kathmandu, I overlaid Shipton's new scene on to the day in the high snows—filled out by Sen Tensing's vivid tale told in the dark. His non-scientific mind giving the footprints animation, filling fears with form. With the context thusly painted, the Yeti has gained a physical description which, in many ways more vividly described the animal than any packaging in the bazaar—a description of words not presented as facts, yet holding validity because the speaker is a local authority.

Shipton himself remains tight to the facts. But what he gives now, beyond the earlier one photograph and a mountaineer's route quest, is folklore brilliant in sober English reporting, a bit understated yet vividly rich in its attribution from local culture.

☙

IN 1971, CRONIN AND MCNEELY'S EXPEDITION HEADED TO THE BARUN. That year I returned to America. Seeking funding for their work, McNeely and Cronin were open about their intent to look for Yetis, though their stated mission was an ecological description of eastern Nepal.

But with the whole of Nepal as a possible research area, they went where Slick's second expedition had made its most suggestive finds, the Arun and Barun valley system. For two years, working out of a base in the heart of the prime Yeti country, they hoped that while learning about the habitat, the Yeti would find them. In January 1973, the Associated Press reported that their expedition had found tracks high on a ridge in the snow. Plaster casts were made. From three column inches in the *Baltimore Sun* (no more in the *New York Times*), I inferred that the tracks fit the Tombazi pattern.

Grandpa now lived with my parents, having left the Indian jungles after sixty years, because my father had joined the faculty at Johns Hopkins to found the Department of International Health. When I passed through Baltimore while on my way to my home now in the West Virginia mountains, Grandpa and I talked. One day, shyly as though afraid he might shock me, he asked, 'Danny, have you ever wondered about the Yeti? There are things in the jungle we do not understand. Cronin and McNeely may have something.' Grandpa and I talked into the night about how their strategy might be improved.

'They're doing the basic part right,' Grandpa said. 'They're sitting in the best place they know, letting the Yeti come to them. You know, it takes just one animal to prove that the Yeti exists.'

'I wonder what the whole trail looked like,' I replied. 'What did they really find? Someday, Grandpa, I shall go and look.' On 13 December 1973, Grandpa died. Neither he nor I had yet seen the plaster casts or the photographs. I held on to the idea—researchers kept reporting individual prints; might there be some answer in the trail?

Then in 1979, forty-two years after his first Yeti discovery, Lord Hunt, who led the 1952 British expedition that first climbed Everest, found astounding tracks nearly 2 inches longer than Shipton's while on an excursion in the Himalaya. Returning to London, he proclaimed to a world that had more or less lost interest that the real question was not whether Yetis existed, but how long they would continue to evade attempts for their discovery.

The same year John Whyte presented a report on BBC with photographs of tracks, smaller than Lord Hunt's. These fit Tombazi's measurements, roughly 8-by-4 inches, with four toes and a thumb-like inside digit. Was Slick right? Were there two Yetis? Or did Whyte's prints belong to a juvenile? Whyte added the dimension of sound. From a cliff above, while photographing the tracks, expedition members heard a piercing scream that lasted about ten seconds, a scream their sherpas identified as the Yeti's.

Shown on television, Whyte's tracks and Lord Hunt's assertion rekindled the British interest in the Yeti, and after almost ten years of nothing suddenly discoveries ensued. In 1980, a Polish Everest expedition found tracks—the large, now fourteen-inch variety. Then in 1986, near my home valleys of the Indian Himalaya, the British traveller Anthony B. Wooldridge came upon a Yeti standing in a gully. He watched it for forty-five minutes and took photographs. Wooldridge obligingly sent me copies. At last we knew what the Yeti looked like after eighty-seven years of searching since Wadell's sighting. Laboratories and the most conservative zoologists double-checked his photography. Nothing suggested a hoax. Here is his report from the 1986 volume of the *Interdisciplinary Journal of the International Society of Cryptozoology*.

> At the point where a set of tracks led off across the slope behind and beyond a spindly shrub ... was a large, erect shape perhaps up to 2 meters tall. ... It was difficult to restrain my excitement as I came to the realization that the only animal I could think of which remotely resembled this one before me was the yeti. ... It was standing with its legs apart, apparently looking down the slope, with its right shoulder turned toward me. The head was large and squarish, and the whole body appeared to be covered with dark hair, although the upper arm was a slightly lighter color. ... I took a number of photographs.[10]

When Wooldridge made his discovery, I wrote to him, and he sent me a complete report, one aspect of which particularly puzzled me: his animal stood still in the open for forty-five minutes—and all reports suggest Yetis are extremely shy. Wooldridge thought his animal was dazed by a fall, but a dazed mammal, I knew, will lower its head below its heart to improve circulation. Might, I wondered, an animal that is behaving totally out of character in reality not be an animal at all?

There was, of course, the other explanation. Wooldridge could have been hypoxic from exertion and altitude; possibly also mildly hypothermic as he reports he'd been running in wet snow. So with nothing but a Yeti-looking shape to suggest its reality, I guessed what he had been photographing was a rock or a tree stump. I wrote to Wooldridge saying so, and he wrote back and then later published the same, following a return visit to the site, concurring that he'd mistaken a rock for a living creature.

Notes

1. Jim Corbett, *Man-Eaters of Kumaon* (New Delhi: Oxford University Press, 1944), pp. 81–2.

2. Corbett, *Man-Eaters of Kumaon*, pp. 86–7.
3. Corbett, *Man-Eaters of Kumaon*, pp. 44–5.
4. Peter Byrne, unpublished documents, quoted in Loren Coleman, *Tom Slick and the Search for the Yeti* (Boston: Faber & Faber, 1989), pp. 62–3.
5. Foreign Service Despatch, 30 November 1959, 'Regulations Governing Mountain Climbing Expeditions in Nepal—Relating to the Yeti', signed by Counsellor Ernest H. Fisk, American Embassy, Kathmandu, Nepal (quoted on Slate.com/blogs/the_vault/2103/02/26).
6. Sir Edmund Hillary, *View from the Summit* (London: Doubleday/ Corgi Books, 2000), p. 242.
7. *The World Book Encyclopedia 1961: Annual Supplement*, S.V.E. Hillary (Chicago: Field Enterprises).
8. William C. Osman Hill, 1961, 'Abominable Snowmen: The Present Position', *Oryx*, VI(2): 86–98.
9. Eric Shipton, *The Untravelled World: An Autobiography* (New York: Charles Scribner's & Sons, 1969), pp. 195–7.
10. Anthony Wooldridge, 1987, 'Yeti Discovery in Western Himalayas', *International Journal of the International Society of Cryptozoology*, Vol. 6, pp. 145–6.

six

Footprints Melting into Rivers

Without exception, all Yeti evidence to suggest an authentic animal—notwithstanding the fun stone footprint found on entering the deep jungle of the Barun Valley and other ciphers like it through the Everest region—is found in the snow. And when those snows melt, that evidence disappears.

Himalayan snows that tower almost beyond reach seem to be secure. Their glaciers endure for centuries, seemingly set apart halfway to outer space. It would seem likely that footprints in these would be preserved. But in reality, the footprints start disappearing the instant they are made, sublimating into the air because air pressure at altitudes where glaciers endure is half that of sea level and because of the doubly intense high-altitude sun.

The river that gathers the waters of every glacier of eastern Nepal, and hence all evidence of the Yeti's reality, is the Sun Kosi. From snows on Everest and her sisters, Makalu, Lhotse, Cho Oyu, Shishapagma, and thirty others over 20,000 feet, glacial trickles gather into streams, then tumble in the tumult of the Sun Kosi. In October 1970, four of us went to ride these waters.

Our descent was the first, succeeding where others had died. A year before, Sir Edmund Hillary tried to ascend this river in a jetboat. (The first passage, he thought, would be safer upstream, allowing him the option of turning around.) There is a singular calling in doing what has not been done before; primal vibrancy comes to life with Nature met red in tooth and claw. The beguiling seduction of attempting 'a first' is amplified by the immediacy of living felt in every moment as to whether you'll even return home.

It is the attempt to try something that challenges a person, for the unknown of the event is linked with the unknown capacity of oneself. As Hillary shared with me years after he and Tenzing did their great climb, it is not a conquest but an inquest into possibilities. Footprints on places that did not have them before open a trail into the self, and in going through and coming out, life opens, questing inward.

Our expedition that opened the Sun Kosi River (also the first successful descent of any major Himalayan river) initiated what is now, four decades later, a new Himalayan sport: river running. Where previously adventure in the Himalaya had been about going up, now available is the challenge of going down, allowing thousands now to ride the Himalaya, just as hundreds every year climb Everest since Tenzing and Hillary opened the great summit. So to discover the currents of that which is uncontrolled, we thrust ourselves into the Sun Kosi's waters.

ତ୍ୟ

IN MOUNTAINS LESS STEEP THAN THE HIMALAYA, along riverbanks are where peoples' paths often first extend, the riverine grass gets padded down, a dirt track carved, and in time maybe a road penetrates. Flowing then through the valleys is the commerce of domesticating the mountain.

But in the world's highest mountains, smoothed river grades do not offer that opportunity. In the Himalaya escarpments

climb out of a river at its every turn. Rivers are barriers to travelling. Lands where banks would typically be found are cliffs.
Trails people might etch to use the river's grade obliterate with
the next flood, as in these vertical valleys gathering waters tumbling, water levels surging, water forces amplifying, rising rock
becomes reshaped.

The trembling rock alongside these rivers that comes from
this intense cascade rises from the deep origins of this land. Our
planet, seemingly rock hard, so presses against itself that the rock
inside it bubbles. In pulling together, the inner rock of the Earth
itself moves in rivers, and that molten movement one hundred
million years ago parted the protocontinent Gondwana. This was
in the age of algae, when proto life also began oozing out of the
waters on to the earth. Gondwana's fragments migrated, making
new places—Africa, Antarctica, Australia, the Americas—but
one piece, the Indian Subcontinent, slid northeast through the
Tethys Sea towards union with the other great protocontinent,
Eurasia. If the Himalaya today tremble such that they quake
with the mountains still rising, contemplate the quaking then as
earthen plates moved, being driven by boiling stone in the heart
of the planet below, through the waters of the Tethys Sea at a
shake-the-world speed four times faster than that at which human
fingernails grow.

So the Indian fragment collided with Eurasia—and the
Himalaya started rising on what were then flat Tethyan shores;
the planet giving birth out of herself, lifting up an 1,800-mile
swath of rising rock, oceanic floor rising into the sky. This, which
is the highest wreckage our planet has produced, continues even
today. The Gondwanan fragment continues to burrow, now
10 miles down into the still boiling magma, and the rivers of rock
inside continue to elevate the Tibetan Plateau (the size of western
Europe), lifting it already 4 miles above the sea and slicing into
the Earth yawning chasms on the other side of this lift such as the
world's deepest lake, Baikal.

At the 1,800-mile separated ends of this planetary collision are embracing lines of water—the rivers Indus and Brahmaputra encircling this supercontinental creation. Both start as trickles from snow on the sacred mountain Kailash in Tibet. The Indus flows north and then turns west to loop the northwest end of the Himalaya. From the same mountain the Brahmaputra flows east and then south, looping the other. And near the base of this mountain another river also rises, the Ganges, slicing the Himalaya to drain the mountain range's southern flank.

As the Ganges collects the southern rivers, into it gushes the Sun Kosi. Their waters continue across the Indian plains, nourishing the fertility of India until the remarkable happens. The rivers Ganges (which is female) and Brahmaputra (which is male) have a union again—waters that started in separate directions on the same mountain, one east and the other south, unite in their ending. Jointly they travel their last miles, and then enter the Indian Ocean, the old Tethys Sea. Waters from one mountain encircle the greatest fence on earth, and then flow back together.

With the circle of these waters is another circle, not across the Earth but, like the mountains they encompass, circling to the sky. As riverine waters flow down, vapourized moisture ascends, lifted from the sea to the sky where the water turns to snow. ('Himalaya' in Sanskrit means 'abode of snows'.) Pulled back down by the earth, snow compresses to glaciers, and the hard ice slides down as the rocks underneath move up. On these moving glaciers, some mystery of animate life makes footprints. When the moving snows with their prints arrive at the end of their journey, they melt, and the footprints of mystery join the flow.

Out of the glaciers of Shisha Pagma, the earth's fourteenth highest summit, comes the Sun Kosi (Golden River). Crossing from China into Nepal, other rivers join: Tamba Kosi (Copper River), Bhote Kosi (Tibetan River), Dudh Kosi (Milky River), the Arun, and the Tamur. Where these rivers begin are the slopes of the world's sixth, fifth, fourth, third, and first highest summits.

6.2   The Lingum Shrine at the Snowfield Where the Arun/Barun
Rivers Begin

Source: Author

Powered by their joined waters, their force breaks through the final
Himalayan barrier, the Chotra Gorge of the Mahabharat Lekh,
the last great upswelled rib of mountains.

ᏩᎤ

FOUR OF US WENT TO THE RIVER: Terry, Cherie, Carl, and
I. We thought we had two advantages. First, I had flown over the
river in a helicopter, so I knew there was no great waterfall. After

Sir Edmund had a member of his team killed, questions had grown among us who lived in Kathmandu regarding whether the river was runnable. Second, we had a special boat. Earlier rafts navigated using paddles. But the boat we acquired had oars, a steering oar on the stern and a pair in the centre to propel and turn—then with the two inside having paddles, our crew could spin that raft. And there was a third special feature that gave boldness—we were between the ages of sixteen and twenty-six.

Arriving at the banks, apart from the currents, our lives stood in a circular whole. Attention could roam and choose through 360 degree perceptions. But when pushing into the river, the river took hold and we were directed by the force. We cascaded down a vibrating line. A once-circle of options became a defining line. Driven down the line we knew only upstream and had just one future: down, bringing forward life questions about what was before not knowing. If we willed, we could stop here, there, seemingly anywhere. But such were just pauses on the line we tumbled of questions on the until-then-not-solved challenge.

We put the raft into the Tamba Kosi just south of the Tibetan border. Our descent lasted four days and nights. The Sun Kosi system crosses Nepal from the north at the Chinese border to the south as it flows into the Ganges—it also crosses half the country from west to east, carrying a third of the country's waters.

Two hundred miles were lived. The first day we lost altitude fast; a rapid every eight minutes. The difficult day was the second, with two rapids we ran just right—for on one rapid the side of the river we bolted down, we avoided a jagged cliff that could have shredded the rubber tubes, and through the other rapid we passed a vacuous hydraulic hole where the river swirled into itself. The third day we rode a memorable half-hour of consistent froth.

Our ride was at the end of the monsoon when currents run full, ebbing from the flood. This higher water smothered many boulders, which in planning the trip we worried might destroy us, so we opted for the greater water of post-monsoon. We were aware one could be thrown out, as had been the fate of one of Hillary's team. But, we wanted the extra water to lift us above the boulders, though we knew that the water was expanding the hydraulic holes to greater size. (Current trips down, now that the rapids and boulders are known, wisely avoid such high water.)

What came with that high water was exhilarating speed. Once we watched a tree riding ahead of us, it was just the massive trunk, its branches torn from it in rapids above, as the tree had been torn from slopes in the jungle above. This tree held to the centre of the current as we stayed on the less turbulent river edge. Bobbing along beside us for a while, the log then abruptly got sucked into one of those hydraulic holes, a moment later catapulted out. We let it ride on ahead. Our air-filled raft floated with buoyancy a log did not have, and it had energetic oars to swing us around the hydraulic hole's rim. Sometimes it is best to let life pass.

Today, with the river known, in guided trips down the Sun Kosi, tourists are often told by their guides to look for the Yeti in the patches of forest alongside. All river trips are unique voyages, no matter the number of times you take them. The river, every river, is new every day. And in the continual new discovery of this wild, it is a valid urge each of us has to discover the unknown of the way as well as our unknown within.

Today our species pounds forward through human-induced cataracts of a changing climate and a diminishing diversity of species. Our journey brings an ending of the once wild. We have climbed the highest peaks, explored every land, and tamed so many dangers. Few firsts remain—except the unknown journey forward. People ride on a changed planet, not that the planet has changed but rather our understanding of it. The discoveries now are made alongside our journey, not so much in the originality of journeys. The term 'Anthropocene' is now used to describe this new human-made age. What we have shaped is more than a warmer climate and a loss of species, a readjusting of the older flow of Nature. One consequence of this is arrogance that Nature holds us less powerfully now, in hubris making us tremble less with its forces.

The human uses of our many energies may have caused the new planetary visage, but we cannot control the age. Time courses on. We must learn to ride it like a river. (Lest we not, cataclysms wake us up.) Whether we fully intended to enter this voyage is no longer open to choice. We ride forces of our shaping—it may help a bit to have technology to keep us out of holes, like a satellite or helicopter view and a conveyance with powerful oars—but we have created and are embarked on a new first descent, one that carries us not down rivers on the planet but through the reshaping of planetary, indeed existential, systems.

We do not know the larger river of Life we now run. The earlier conceit that initiated the global changes races into currents beyond understanding. What carries us on into the unknown is that we cannot go back. Time, like all rivers, runs in one direction, for even if there is a tide that appears to push the river back, the river always runs on.

On this Sun Kosi exploration, we were free in so many ways—free from an engagement with the quest, then free from the river when the four of us were standing on its banks talking. We believed the river, like life, was to be a grand adventure. To assay ourselves we then pushed in for testing. But once in, freedom was no longer there. We were clutched and carried. And we learnt to live with it so that we could go with it. Cascading down, no longer able to turn away, we were harnessed into this flow, like that log.

In these rivers boulders, like us, also run with the current. They are bottom movers, but still pushed like us by the descending water. The boulders create the rapids we coursed over. Sometimes house-sized, the rocks shift rapids as they roll—and at the end of the monsoon the boulders were rolling. Thuds in waters below echoed in the raft's rubber tubes. We could feel them. Rivers move mountains and within them mountains move.

Hearing the rapids ahead, our oars strained, a brief moment to poise us in the hoped-for right place, the river accelerating as it approached the sudden descent ahead. Entering the rapids we spun the prow, sped up in the accelerating current by driving the oars to have direction in the desired flow. Without missing a splash the current drove us on.

Eddies, rocks, fields, jungles, spirits, and people; a montage of life passing, the river tying all together—the rhythm of rapids punctu-ating our lives: forward, downwards, made strong by rain and melted snows. Our destiny was to flow with the push of water that came from the sea, coursing us back towards the sea, water having circled to the sky now heading home. We did not choose our direction and only minutely adjusted our speed. Being inside this movement felt as though inside the arteries of the changing planet.

Our future loomed with questions. Pounded by rapids, the pres-ent cascaded with answers, answers that floated into memories in the pools that followed, answers that slipped through our fingers, leaving them wet as fingers gripped and paddled. Carried out of one event, passed to another, the actions by us were to glorify and to enjoy that downward rush of our living. Each splash whetted the way to the next venture.

Early on in the second day we passed the furthest point where an earlier expedition of a friend had stopped. As best as we could learn before our trip, navigating the whole Sun Kosi had been attempted four times. One had been by this friend; he had described the rapids where they lost their boat. Instead of attempting to run the rapids through the middle as they had done, getting themselves drawn back and spun in with one of those hydraulic holes, our route slipped over on the left. After the tumult of those rapids, they had pulled out and walked a slow week back to Kathmandu, happy to be alive. (Another part of their problem was their too-small raft.)

So knowing about those four attempts, we thought now from that point we were the first descending these new waters. Each rapid fresh with uncertainty, the big unknown came closer, the mystery-filled Chotra Gorge. What would be the size of its waters through that final gap in the Himalaya? While the river felt powerful now (our Sun Kosi having gathered the Tamba, Bhote Kosi, and Dudh Kosi), when we entered that gorge, the then-larger current would be joined by the Arun, the oldest river of the Himalaya, and also merged with the Tamur; six rivers running then as one.

With everlasting certainty the current drove on, the unknown was not yet in our present, constantly growing from gathering the streams we passed. The current seemed to pull us towards the boulders, but we kept pivoting in the waters and kept on going. And in those quieter times we floated amid beauty that defined our lives: a strong blue sky; a hard, hot sun on green growing rice. As the river carried us to lower elevations, we passed fields with ripening rice and saw people harvesting it. Towering cliffs were behind us, and the river was entering the lands of people. The people were working. Waving, sometimes shouting to others nearby and pointing towards us.

The sun scorched our skin. Our sizzled bodies slipped overboard, and the tingling skin melted the burn with the chill. While the raft went down, at times now we floated beside, letting ourselves be carried by the river, made into a particle in the current. Or sometimes we just sat there in the boat, trying to be apart from the river, waiting.

Rapids give news of their arrival first by sound. A growl ... then that grows. Approaching a rapids, it first shows as a line across the river. The line of river going down has suddenly a line across. The growl then roars; the lower jaw of the river falls open. Into it our raft then went, throwing the weight of the boat back to land prow high, stroking with the oars to pull the raft forward from backwashing into the chomp of the water's tumbling teeth.

From Nepal's central hills we drifted into her sparsely settled foothills. Rounding a bend, we entered a sylvan pocket cut off by cliffs all around, a primeval bank where we pulled ashore. Wild jungles stood, seemingly a protected side-valley holding no people. But people had indeed come a stump cut long ago assured us. In one clump of trees were twin banana trees sometime somehow planted with bananas then ripe, fruits for us and fresh rest for our bodies. We lay on the moss, wet and cool and thick.

Back in, in front of our ride, standing up to its knees on a rock in the water and pumping its tail, a white-capped river chat dipped its beak and trilled out 'SHREE'. Around the next bend, across the river's span stretched a spider's web. Its lattice of threads danced with drops scattering light with rainbows. More droplets splashed on to it the instant we approached, quivering on the strands, star prisms sparkling before a blue sky beyond. The current, though, crashed us through, driving us on.

Traditional river navigation in the Himalaya consists of crossings by dugout canoes at pools between rapids. The dugout is towed upstream to the foot of the rapids above. Passengers, perhaps a goat, maybe some chickens, snuggle in. Usually, neither the crew nor the passengers can swim. With gunnels inches above water and passenger fingers clenching the sides, a crew member in the bow and his partner in the stern push off, they paddle vigorously for the opposite bank before the current sucks them into rapids approaching from below.

A lone Nepali waved, calling us to his shore. As we talked, he was, he said, another like us, who had tried using the river to travel. Facing costs of his daughter's wedding, he harvested a clump of valuable cane, lashed it into a bundle 10 feet across, creating a giant, man-made, floating island that tumbled and rolled as he— clambering to always stay on top—travelled down. He rode his bundle through the Chotra Gorge, across the border into India. 'I got my money, but never,' he vowed, 'never will I do that again.' So we smiled getting back into our rubber hope. At least from here we are not the first going down.

Waves. Waterfalls. Rocks. Our souls crowded with personal thoughts, plans were remembered, prompting private hopes to be renewed for our life hereafter. At first on this expedition, we tried to protect ourselves from being too close to each other, to maintain a distance as the four of us worked out mutual accommodations. There is pride in trying something never done before. It holds a group together, but there are also the issues that separate. Yet the challenges accomplished brought our group together, for alone Terry, Cherie, Carl, and I could not have made the raft spin and accelerate. In working together, a wrong move by one could flip or sink the rest. In pulling the oars we came closer. Entering the expedition, we had entrusted our lives to each other. Going forward together, when once I was thrown out of the raft, hands from these friends gripped my wrist, tendons bulged, and I was pulled back up over the black tubes into the shared circle.

Then, as the river kept flowing on, our trip was over. The Chotra Gorge, filled by the body of six rivers, turned out to be fast and smooth for our passage. As we exited the gorge, the Indian border appeared ahead. Rivers do not need passports, but people do. Now we stood on a different type of shore where on the plains of India, rivers had banks. Rivers were wide and they ran slowly. It was the same river, of course, but the shore we had reached was of completion. No longer would we run with the river that had been a part of our lives. The river would now run in us. And from this running we came to know that hidden behind the ranges was a call to go back into the mountains.

seven

Towards the Barun Jungles

Previous page:

7.1 Cartoon

Source: Dan Piraro

May 1979 was three years still before we entered the Barun, and my Yeti search had already been underway for twenty-three years. The king's advice to explore the Barun was to come the day after the lunch I was now headed to with Dan Terry, a companion from school days. We were walking up the path to Bob and Linda Fleming's small bungalow in Kathmandu, when ...

'EEAL!' Terry and I turned.

Against the brick wall, half hidden amid ferns, was a six-feet-long grey Nepali rat snake of the genus *Ptyas*. Disappearing into its mouth, its body pierced by the snake's razor-sharp teeth, was an ordinary frog of the genus *Rana*; the frog until a moment before was a resident of the nearby rice paddy. It kicked its legs. The squeal lingered, as Bob opened his bungalow door. On the coffee table waited a tray of teacups, teapot, strainer, milk, sugar, and glucose biscuits.

Picking up the teacup Bob just handed me after he'd added a dollop of buffalo cream, I mentioned the Marsyandi Gorge from where Terry and I had just come, specifically the jungle where the gorge narrows between the Annapurna range and the Manaslu massif. I'd heard of this jungle from the traders who first shared their bun manchi stories eighteen years before. So, on this trip when we headed to the Manang Valley on a medical research expedition, I stopped in the village, telling Bob now how the villagers believe jungle men come into their fields at night, and in defence they build fences around their fields.

Bob knew how the trail there runs on the side of the gorge, mentioned the thin black cliff above, with a jungle all the way to the trail, and spoke of how the place gets extra moisture as the weather rushes through between the mountain massifs of the Annapurnas and Manaslu. He agreed that it was biologically rich, but he was dismissive of the possibility of finding unknown animals there given the limited area of the jungle, and he was certain that what invaded the fields was a monkey or a bear.

I was not persuaded. People don't build fences because of their imagination. Bamboo fences would not stop bears, and monkeys would hop over.

Terry pointed to the obvious fact that villagers knew monkeys and bears. So why make up the bun manchi when they recognize bear and monkey signs? Terry and I were cautious, for through our school years Bob's knowledge of Himalayan natural history had us in awe. After finishing his PhD, he had now spent fifteen years exploring the Himalaya nearly full-time.

Bob tweaked our relentless persistence, 'You two have looked for Yetis even inside cars.' He was reminding us of how in 1968, we had a VW van that we drove from Germany to India and had named the bus 'Yeti ka Bhai' (Yeti's brother).

It was the summer after my first year in graduate school. Three days before our trip Terry sent a message to my Harvard mail slot: 'You buy VW van in Germany. Rendezvous June 1st Switzerland. I pay gas to India.' The year 1968 was when the Beatles were riding their yellow submarine, and hippies were headed to Afghan hashish and Indian religions. Yeti searches made sense in that era. But

Terry's cryptic message did not say which airport in Switzerland. So on 1 June, having in my possession a USD 400 WV van, I met all the flights from the US in Zurich and, via long-distance phone, paged all the flights arriving in Geneva. Terry paid the phone bill as well as the gas bill.

We sold seats by the mile to hippies. Adjusting the clutch under Yeti ka Bhai in Ankara, Turkey, we learnt that Bobby Kennedy had been shot. South of Mount Ararat we dodged bandits. In Afghanistan's Baluchistan Desert, with the temperature a blistering 121 degrees Fahrenheit in the shade, our crankshaft bearings burned out. Five days later, we were driving on bearings laminated up from a Spam can and Harvard stationery. The bearings blew again, this time with the engine entrails flying through the crankcase. When we arrived in India, Bob laughed seeing the blue and white Yeti ka Bhai. Many would have called it an abominable bus.

I was half expecting Bob now to make another joke. He set down his teacup, leaned in, and dropped his voice, 'I'm not putting you guys off. I doubt the Yeti is bun manchi, or a hominoid, but six years ago McNeely and Cronin showed me a plaster cast of the footprint they found. It was made by no Himalayan animal I know, was similar to a gorilla's footprint with a primate-like thumb. McNeely and Cronin's other field discoveries have held, like their honeyguide, which was then a new bird for Nepal, so I doubt this is a hoax.'

Bob shifted in his chair and, in the discomfort of the moment, suggested that we move to the lunch table. As we resettled around plates of entwined spaghetti, no one said anything. The evidence was before us. Bob, if not a Yeti believer, seemed at least to be a puzzler, and appeared unsure about how to state his position. Then he said what we all knew: something was making footprints no one could explain.

None of us responded. Bob was seated at the head of the table; Linda, at the other end. She smiled as Terry and I focused on noodles and sauce. Let Bob work himself in deeper, we thought. Bob could not explain the photographed footprints and was also puzzled by a plaster cast he has held.

Bob carefully separated the issues. He pushed aside allegations of the bun manchi, which he believed to be a monkey or a bear, noting that their prints could appear hominoid. He believed that Cronin and McNeely's findings as scientists were reliable. Some real animal visited their 12,000-foot camp on that high Barun ridge. Outside their tent were tracks that did not belong to any animal he knew.

Like a spy who had held his secret too long, Bob kept talking. Terry and I remained bent over our plates. The Cronin and McNeely evidence was not the news. The news was that our authority on natural history since our schooldays accepted this evidence. He was not offering a conclusion, but uncertainty that had baffled not just him but also George Schaller. George and Bob had talked at this dinner table. George was the one who said that the prints were similar to a mountain gorilla's, and George knew large mammals, having knocked off definitive studies of the mountain gorilla, lion, and tiger. He had come to Nepal to add to that list a quest for the snow leopard.

Dessert then was brought to the table: chilled litchis, the insider's fruit of the tropics. Beneath a gnarled, brown, thin skin is a dew-like flesh wrapped a marble-sized hard seed. Squeezing on the skin to eject the fruit that is inside, I popped one litchi after another into my mouth. Delicate and moist, they melted on my tongue, like morning mist disappearing at dawn. Terry and I remained in listening mode filling our mouths with litchis.

Back to the living room and another pot of tea. Whatever this some 'thing' was, Bob believed that it did not live in the snows where its evidence had been found, agreeing with William C. Osman Hill. The animal was likely to be a herbivore; Bob gave an interesting reason that herbivores could be reclusive because carnivores had larger home ranges and so they signalled their presence with more discoverable kills. To support its size, a large herbivore, Yeti, gorilla, or panda would need a lot of grass or bamboo, and that meant this snowman must live in high grasses and bamboo.

Bob was careful not to step into the tempting area of seriously proposing a new animal for the footprints. He was quite sure the mystery would be explained by a new way of understanding a known animal's footprints. Terry was convinced about the bamboo, noting that bamboo forests are great hiding places. (He and I had once been entangled for nearly a day in such a miserable thicket in the Kullu Valley.) He thought that while the animal may eat bamboo, it would not live in that world, proposing that we should examine caves.

Bob then raised a point I had never seen discussed: the timing of the Yeti searches. Footprints had been usually found in the spring, sometimes fall. In those seasons animals are generally on the move, and perhaps that is why the footprints are found on high glaciers, as the 'thing' moves from its dense habitat and passes from valley to valley before or after winter. When, though he wondered, would be the best time to find the animal?

We all agreed that summer would be bad as monsoon and leeches would overwhelm. Terry suggested spring as in an open jungle it would be possible to see further. I opted for fall when food supply would be greatest and animals are sexually active— hopefully vocally active. Caves gave me little hope because for years I had been trying that.

Bob reminded us of winter, an idea neither Terry nor I liked; both of us having had some of our most unpleasant days in the jungle with low snow and winter rains. But Bob suggested that was the reason—winter drives animals lower, and with the snow we could exactly the right medium to connect to the mysterious footprints. Moreover, whatever and wherever the mystery was, we should assume the individuals were not many, and as there is now little pristine habitat low in the Himalaya, in winter those few animals would concentrate in that remaining jungle. With snow, it would be like having the animals drop off their business cards where hopefully some would carry the distinctive imprint.

Leaving the bungalow, we looked in the direction of the brick wall where now rising from amid the ferns was not one but two rat snakes twisted around each other like two thick vines. They were in a fight for territory. In Nepali folklore, the snake is the lord of the underworld. Two rat snakes swaying like the ones before us are periodically mistaken for a cobra, the particular incarnation of the lord of the underworld. As we watched, the smaller snake seemed to prevail, and the other with what appeared to be a midriff bulge slipped under the wall.

The following evening, leaving Kathmandu, I sat on the right side of the Royal Nepal Airlines jet as it climbed through billowing clouds. Earlier that day I had met the king. We had a custom to meet on my departure every time I visited his country to share what I had found on that visit. And that day, after sharing our survey in the Manang Valley, we talked about my plan to head out again in search of the footprint-maker. Shah Dev said then what he had said before: 'The Barun is the densest jungle in my kingdom.'

As the jet ascended through the clouds, outside the plane, higher still above, were Himalayan snow and ice, land whose abode was in the clouds. The flight passed over the Marsyandi valley where the villagers were so worried about the bun manchi, leaving the great Manaslu massif behind, bringing up the Annapurnas— ice and rocks out of the window, amid the clouds.

As the plane levelled, colours on the snows changed. The sun dropped off the horizon, reddening like a fireball as it fell, shining across what minutes before were white clouds, firing them into reds. As it set further, the snows, also white earlier, turned lavender, flashing to electric red as the sun found a crack in the clouds. As we flew on, peaks turned pink, and the sun, now behind the earth, reflected on to the summits from the outer atmosphere. When the Royal Nepal Airlines leaves Kathmandu on time, a seat on the right side of the Kathmandu-to-Delhi evening flight is one of the great aeroplane rides over the earth.

❧

A YEAR LATER, I WAS IN MY MOUNTAINTOP HOME. A fire was burning in the fireplace, and I sat with the just-off-the-press book, *The Arun: A Natural History of the World's Deepest Valley*. Presented were Cronin and McNeely's discoveries. While serving as Peace Corps volunteers in Thailand, they had become interested in the Yeti, but the expedition they designed had a broad ecological scope; but with some intentionality siting it in the Arun Valley where Tom Slick had made his most enigmatic Yeti finds, specifically in the valley immediately south of the Barun.

The book opens describing the trek to their base camp. It began at Tumlingtar where a short takeoff-and-landing airstrip happens also to be situated at the crossing of foot trails that connect Nepal east–west and north–south. Cronin's mention of Tumlingtar brought back my years in Nepal's family-planning programme, specifically the 1969 Tumlingtar vasectomy camp. A young physician decided to demonstrate vasectomies to quell local fears about our upcoming campaign. Unfortunately, he cut and tied his patient's testicular vein instead of the vas deferens. The villager walked home. The next day the man was carried back crying with pain, his scrotum ballooned to the size of a volleyball. The story flashed from Tumlingtar, porter-to-porter, tea stall-to-tea stall, exaggerated by some to castration. Throughout the kingdom strong men became terrified.

Khandbari Bazaar is the big town on their route. I helped set up a family-planning office there. On Wednesdays, the day of the weekly market, people milled about having come in from the villages—women of that region dressed in especially bright colours. At the market I paid a paltry seventy rupees for 50 pounds of the tastiest tangerines, roped the basket to the outside of the helicopter that took me back to Kathmandu. That was the trip when I surveyed the Sun Kosi River for our successful river run.

In the jungles to the south of the Barun, Cronin's biological story unfolds. The eighth chapter introduces the Yeti. Cronin presents a veteran hunter who says: 'Oh, yes! We have many kinds of wild animals in these forests. There are bears, and musk deer, and Yeti, and pandas, and leopards, and civets, and monkeys, and many, many more.'[1]

It was so matter of fact. The Yeti is a distinct animal. A shikari, such as the man speaking, is an interesting authority, and perhaps very accurate. Village hunters' livelihoods depend on jungle knowledge—not where to bag trophies, knowing what animals can be found where and when, specific knowledge that will provide food. Skills are learnt from one's father, and often his father's father, a circle of family secrets—stories about a tree, changes of a stream, populations of a bird. In most villages, only a few families are hunters.

One skill every shikari has is of storytelling. Hunters brag about what they have done, and share little about their next hunt. Each hunter has his own style. Some exaggerate, others understate, and neither style can be trusted when possible employment is involved. Cronin's hunter seems to be enjoying his story and, I assume, trying to get employment.

But this man is never mentioned again. After a Yeti revelation, why let this guy walk right out of the camp if the information was credible enough to lead the chapter? Why not pay him to show the Yeti's habitat? Nothing more is said about him. Hunters often double as porters when they can get the pay of a Western expedition. This expedition's area straddled the trail to Makalu. The hunter could have known of Western interest in the Yeti from having worked for mountaineers. He may have worked on Tom Slick's expeditions fifteen years before when they hired hundreds of local people.

Or maybe Cronin didn't really understand. His book never reveals how fluent any member of the team was in Nepali. Bob said that as ex-Peace Corps from Thailand, they got along in Nepali.

But indications in the book were that their members with dual fluency were the Sherpa guides. Maybe the use of 'Yeti' here came from the Sherpas. If a local villager mentioned the bun manchi, might not a Sherpa translate 'jungle man' as 'Yeti'? What word did the hunter really use? In my time in Tumlingtar and Khandbari only two years before their time, 'Yeti' did not seem to be an Arun Valley word.

Again, it is clear that for Yeti hunting language is an essential skill. Questing for the Yeti appears to be as much about following words as footprints. Cronin uses a word most readers would think is Nepali but is actually Sherpa and now also English. A hunter from these valleys would more naturally have used the words 'shockpa' or 'po gamo', even 'bun manchi'. For my purposes, I must dismiss this testimony and concentrate on the book's photograph that follows, or the cast Bob held.

The fire was burning low now, and in anticipation of the description ahead, I went out to bring in more wood. In a night sharp and pure, my gaze went to the far-away stars. Are humans the only beings? Might some other types be hiding here or beyond? Are animated spirits out amid the stars? Our ancestors believed in prophets who carried in the word. Many trust TV now to bring messages—it shows real pictures. But for mysteries, who informs us today, and how do we know if their messages are credible?

I added new logs to the fire, and charred crusts broke off from the old ones. Fire changes hardened old life into flames. Oxygen, the breath for all life, mixes with relics of life, wood; from that new life seems to burst, dancing, life coming alive again. But this dancing message is life in a different form, bringing forward private thoughts from the finger-like moving signs. Picking up Cronin's book, after presenting the hunter, is a familiar list of Yeti discoveries, then these sentences:

Shortly before dawn the next morning, Howard climbed out of our tent. Immediately, he called excitedly. There, beside the trail we had made to our tents, was a new set of footprints. While we were sleeping, a creature had approached our camp and walked directly between our tents. The Sherpas identified the tracks, without question, as yeti prints. We, without question, were stunned. ...

The prints measured approximately nine inches long by four and thrēe-quarters inches wide. The stride, or distance between the individual prints, was surprisingly short, often less than one foot, and it appeared that the creature has used a slow, cautious walk along this section. The prints showed a short, broad opposable hallux, an asymmetrical arrangement of the toes, and a wide rounded heel. ... Most impressively, their close relationship to Shipton's prints was unmistakable.[2]

February 1981. The United Nations offers me a consultancy in Nepal. My life, though, had expanded—Jennifer and I now have a six-month-old son, Jesse Oak. Taking a baby into the jungle is a way of living I grew up with—my father before me did that and it is how many families continue all over the world. Could our family go into the jungle safely? A special fare from Pan Am airlines allowed Jennifer and Jesse to travel for less than USD 400 extra, giving us a chance to see whether safe travel is possible in the jungle with a baby. After the consultancy, we float down a river on the Nepal–India border in dugout canoes, through gentle

7.2   A Barun Ridge Similar to Where Cronin and McNeely Found Their Footprint

Source: Author

rapids, through drenching rain, and then walk to a 'lost' temple deep in overgrown trees and vines. Here was the octagon well where lore has it that Ram and Sita made love. Adjusting to the special childcare needs in the jungle and surprises on the road, travel rolls at a slower pace. But, as a family we have a vacation, and Jesse stays in good health.

In October 1982, another consultancy provides another windfall. In the years since the tea with the Flemings, while on trips to Kathmandu leading medical expeditions and consultancies, I've ferried out Yeti-chasing supplies. Our expedition that is soon to depart for the Barun Valley, aside from ourselves, will include Bob and Linda, but not Dan Terry. He is working in Afghanistan with his family, a land now under Soviet occupation with work that gives little flexibility.

Before leaving America for the Barun, I survey the Yeti references again. Twenty-three claims must be given some level of credibility. Most are wishful, others string together irrelevant items to arrive at a Yeti option. The descriptions they give of the animal are jumbled: a baggy rear end, a villager's tale of a shaggy fur but clear face, long dangling arms, colour from red to black to brown, a pointed head with a long sloping forehead, bulging eyebrows, feet on backwards, a shuffling gait, and that extraordinary claim of breasts so pendulous that when running they can be slung over the shoulder.

Even after thirty-one years, what continues to attract international attention to this hotchpotch of claims is the 1951 Shipton photograph. But why didn't Shipton and Ward take more photographs that day? The Yeti was a known possibility for both of them; indeed Shipton had encountered such prints earlier. Had they taken a close-up of the other foot more could be deduced. Why not a picture showing the length of the stride? Photographs of other feet would show how the print they chose had been altered by the melting snow. I again compared Shipton's accounts: the first from 1951 that had an offhand mention of the Yeti, and from 1969 where he described every possible detail. And, it is striking that in all of Ward's writing he has not discussed this discovery—for he is a medical scientist.

But more deeply, why are these twenty-three published reports from the high snows automatically interpreted by everyone to indicate a wild hominoid? I, too, do that. Few start asking if a known animal might be making them. I remind myself that in our upcoming expedition I must stay focused on the footprints as it is these alone that are the only trustworthy evidence.

I have become increasingly aware that what is driving this repeating connection to see hominoids is more than the hunger for the 'missing link'. It desires the wild alive in us. It is like the log burning in front of me, which, when ignited, draws viewers with enigmatic flames to ignite our imagination, sparked by what was once life. Hominoid-esque footprints sign a trail that connects to our evolution, projecting a possible relic still of clandestine life in the high valleys. The footprints are real, like a shadow on the land, but animation comes by a hopeful mind. For a hominoid to be successful in hiding, the animal must have intelligence approaching that of humans, and would have to be furry enough to withstand the wet cold at elevations where hair-free people cannot go. The animal could be a gorilla or panda in terms of its body and diet but one that must also have significant intellect to allow it the ability to hide.

In fire is a metaphor: a fire is not really alive. It dances, maybe even talks as fingers of flames sign ideas into the mind, an experience of the not alive though it comes from the once-alive. So it is important not to read too much into either footprints or fires, both of which speak in different ways of life past and 'passed'.

If the Yeti indeed is a hominoid, and not a known animal with mistaken footprints, then the more likely explanation is that instead of a new hominoid it is a half-crazy human living in the wild. Maybe it also wears clothes, and so when seen without its footprints in the snow it looks like normal people. Yetis could be the semi-mad who are cast out from their villages, or simply

7.3   Thangka, a Tibetan-Style Religious Painting, Showing a Yeti Standing on the Mountain to the Extreme Right

Source: Tashi Lama

(A thangka artist, Tashi features later in this narrative as Pasang's son.)

self-inflicted outcasts who walk out of the village to live in the jungles, maybe even still half-supported by their families. Outcasts with huge calloused feet would imprint impressive identity questions in the high snows.

Now, after searching across a quarter of a century and having searched in most of the Himalaya, I know the world's largest mountain range is not untrammelled—I have been into almost every valley system, and where I have not personally gone, Bob Fleming certainly has, being more observant of the wild beings than I. Himalayan valleys are full of modern humans—filled from population growth pressing up from the fertile plains of India and China, a human collective on the planet comprising 40 per cent of humanity.

Yet the footprints remain unexplained. Bob saw casts of those. These are the quests in the Barun Valley we must seek to explain. They were not fakes; they were unlike anything he or George Schaller could explain, with neither Bob nor George thinking the prints to be of a human. So I should must even dismiss the mad outcast option. Bob and George saw greater resemblance to non-human makers. The mystery remains unexplained. Like shadows, footprints are outlines of life; real, yet not real. A hunter does not shoot an animal by shooting its shadow. Our family is not going Yeti hunting—we are going footprint hunting.

7.4   Jesse Being Readied for the Trail

Source: Author

❀

OUR FAMILY APPROACHES THE BARUN. It is still one day's walk away. The discovery of our footprints on the ridge is two weeks away, and three weeks distant is the proposal from Lendoop that the maker of the footprints we found is rukh balu, the tree bear.

With Tumlingtar airstrip three days behind, we look across the Arun Valley; the village of Num lies below on the ridge and Hedangna is across on the opposing ridge. At the head of the valley is Makalu. Snow blows from this fifth-highest summit; a band of clouds beneath it portends the rains of the Barun we shall soon experience. Makalu is notorious for its weather; only four expeditions had, by 1983, made it to its summit. Beyond Makalu stands Lhotse, the fourth highest, and behind her, Everest. 'Earth's sisters in the sky', the Tibetan poet Lhakpa Phuntshok calls them.

With the splendid view in front of us, I swing Jesse off my shoulders. He runs after a cluster of yellow butterflies with black lines in their wings. The air here suddenly smells sweet. I crave for toast and a cup of tea. Just then a porter comes up the trail, carefully placing his feet on each stair-like rock, body bent under two gigantic bundles held by a strap around his forehead.

Cinnamon, that's the smell! He stops at the stone wall where we sit. Releasing the strain off his tumpline, he gives a long whistle—the universal sign throughout Nepal saying that your load is heavy.

I ask, 'Kahan jane ho?' (Where are you going?) The universal greeting of the trails.

'To Hille ... but if prices in Hille are too low, by the big truck to Dharan.'

'Did you cut the cinnamon, or are you carrying it for someone else?'

'It's mine, from the trees in the jungle I know. Slowly, I've gathered it. I do not take too much from one tree because then I cannot go back the next year. Nowadays, some people cut too much, and then the tree dies.'

'Big load you have. It must be two people's load,' I say.

'Three loads.'

It is my turn to whistle—180 pounds!

'Why doesn't one of your children help? Three loads are too much for an old man. Your knees will soon hurt like doors that swing on rusty hinges. And when they do, you will have to stop walking the hills. Three loads are too heavy.'

'Last year one more of my children died. Many children have died; so again I have left my children—if they carry loads, maybe they will also die—what else does a man have as he grows old but his children who walk on after he is gone? So I told them to go to school while I walk the trails. My wife makes sure they go each day while with this load we buy uniforms, shoes, and pencils.'

'What happened last year to your children?' Jennifer asks.

'Last year the sickness came. For six years it had not come. But last year it returned. You know the children's sickness, the one with fever?'

'What does he mean?' Jennifer whispers to me.

'Measles,' I reply quietly. 'Some villagers call it the sickness of children. It sweeps these villages every five to seven years, infects all who have not had it, and then returns when a new group of children are not immune.'

'One son and one daughter are left,' he continues. 'I pray to God that my son does not die, because last year the sickness returned. My son almost died, as did the daughter.'

'If your son has had the children's sickness once, he will never get it again,' I answer.

'Yes. But my wife and I worry … and he might get another sickness. We were thankful this year that our son did not die. But we have lost three daughters. My wife is still strong; soon she will have a baby. Is that your son running over there?'

'That is our son,' I answer. 'When you return home, take your son and daughter to town. A doctor there can protect them from the sickness of children and other sicknesses too. He will give each child an injection—that will help your children even more than school uniforms.'

'Yes, but my children need uniforms for my family to be respected. And, sir, be careful with your son. Why bring such a lovely son to this place? Keep him home as I keep my children.'

'We want our son to see your beautiful home, to learn how to live in these beautiful mountains,' I reply.

The porter is quiet as our group looks across the valley. 'If you think these mountains are beautiful,' he says, 'that is good. But for me these mountains make life hard. You have not had to plough the rocks in my fields. My fields are close to the jungle, and I do not think you have had to protect growing corn from bears.'

'What happened to your three daughters who died?' Jennifer gently asks in Nepali.

'The youngest died last year. We had not given her a name yet; she was called *sanu* (youngest). Both she and my son had fever. The youngest was two and my son was seven.'

Nick had been calculating the cycle, 'Why didn't your son get the fevers six years ago when the children's sickness came before?'

'My wife and I never understood that. Another daughter died, but our son did not. Maybe it was because we sacrificed a chicken at the temple. It is wise to sacrifice chickens when you are worried.'

'The boy was one year old at that time. He would have been nursing and so protected by his mother's immunity,' I quietly insert. 'Also he probably got preferential care because he was a boy, and this recent time with two children sick, the girl was likely ignored.'

Jennifer whispers, 'Dan'l, his family must be doing something wrong. Shouldn't we tell him something?'

'Sir, will you give me permission to leave?' The porter seems uncomfortable. Is it, I wonder, because I've been asking too many questions, or, is it that he knows how many more steps must be taken before he sets down his load in Hille or Dharan?

'Yes, *bistare janos* (walk slowly). With that big load, use a stick when walking downhill; that will save your knees. Purchase rubber-soled sandals when you pass through Khandbari Bazaar. Those sandals will help save your knees—and when your children get the fevers the next time, give them lots of water; boil the water first and add a tiny amount of salt and a little bit of wheat flour to it. Feed that to your children as much as you can.'

'Thank you, sir. When I get to Dharan, I will purchase my items. Then I will have money. With the money I now have I must walk with little food.' He lifts the tumpline from his shoulders and positions it over his eyebrows, bending into the 180-pound load. Two aging legs pump uphill as they have been lifting him for so many years. Cinnamon bundles squeak as they adjust to their own weight; a sweet smell rising, reminding me again of breakfast toast and lingering memories. It will take this man another week to reach Hille. He might have to travel to Dharan, maybe even India. Someplace along the way his money to purchase meals may run out; he'll be walking on the strength of water from the streams. I hope a 'friend' does not get him drunk one night in a border town, for then the bundles may be gone by morning.

Two days later we arrive at the Barun River, meet Lendoop and Myang, and then go into the jungles to discover the footprints on the ridge and the mystery of the scientific explanation for the tree bear.

Notes

1. Edward W. Cronin, *The Arun: A Natural History of the World's Deepest Valley* (Boston: Houghton Mifflin, 1979), p. 153.
2. Cronin, *The Arun*, p. 167.

eight

Our Evidence
Meets Science

Previous page:

8.1 One of My Yak Caravans When Searching the High Snows behind Makalu

Source: Author

March 1983. Footprint and tree-bear discoveries then followed as described earlier. Contrasted with Cronin and McNeely's footprints a decade before, and Shipton and Ward's two decades before that, our footprints came with an explanation with regard to their maker. Bears had been suggested decades before by Evans and Smythe. But now the century-old footprint riddle had an answer as to how those footprints could remain hidden: the snowman lived in trees, and the unknown animal looked like a known animal.

⟋⟍

JENNIFER, JESSE, AND I WAIT AT TUMLINGTAR AIRPORT as the Twin Otter flies in, touches down, long legs galumphing over the field's wallows. The plane turns on the grass field nipped short by goats and water buffalo, and its whining turbines never stop as seventeen passengers disembark and our seventeen board. Turbines roar again, legs chatter again, and this aircraft, specially created for heavy lifting, is airborne.

Baggage fills the aisles bringing the plane's weight to be at full gross generously calculated. Looking down the aisle into the cockpit, as the plane climbs, the Tumlingtar ridge rises in front. Will the pilot circle? No, he pulls the control yoke as we near the ridge, and the plane tail-stands. That sudden pitch causes a radio to slide out of the instrument panel which the pilot catches with his right hand while holding the yoke with his left, all of us then finding what the pilot knows—this side of the ridge has an updraft, and it hefts us over. The co-pilot turns with a wide smile on his face, and our eyes meet as he glances down the aisle; he and the pilot clearly love making their plane dance. He unhooks the latch and shuts the cockpit door.

Peaks rush past. Scratches in the plastic windows cleave the light into a rainbow lattice that makes these snows glitter. On the Kathmandu cocktail circuit, pilots, as they navigate their 'scotch on the rocks', share stories with the attentive who may fly with them the next day, about the 'agriculture' of the Himalayan airspace—how the clouds of these mountains grow both rocks and potatoes.

Flight everywhere follows regimens, the regimens differing whether in New York airspace, or combat. Here in mountains where land can exist above the air, to prevent entering a cloud that may not be just air, the regimen requires turning towards India to work around the cloud. Now, flying over ridge after ridge, our aluminium tube with the updrafts feels as though it is going through the rapids of a river, and shaken loose with that jolting a carry-on bag scoots down the aisle and lodges against a sack of oranges.

As we walk into the Kathmandu airport, Nick jumps out from behind a pillar. 'A tree bear is at the zoo!' he says. Nick left us earlier to go see another part of Nepal while we continued fieldwork in a different part of the Arun Valley. After forty-seven nights in the field, Jennifer and I want news only of a hot shower. Nick continues, 'I went back to the zoo. Remember the bear we saw? I wondered if it could be a tree bear. We don't know how small tree bears are, but that bear is small. The zookeeper said that the bear had been in the zoo for two years; it had not grown. If it is not growing, it could be a tree bear.'

'Yeah, Nick,' I answer, 'but not growing doesn't prove anything. Maybe the bear is a runt. After all, the zoo food must be lousy.'

The next day at the zoo, as I pitch peanuts through the bars of her cage, her mouth is held open inches away. Amid crushed peanuts and saliva, to estimate age I try to fathom the wear on her molars. She loves peanuts, shells, as well as the meats. Would this bear go over the mountain upon smelling peanut butter in the wind? I push a stick through the bars to brush her teeth to better examine her cusps, and her jaws snap. As I work another stick, she curls a front paw around the cage bars. Is her inside digit likely to make a thumb-like print? It holds impressively with an opposable grip. Her breath smells like boiling turnips.

The zookeeper works backwards to figure her age. 'She's been a good bear, never mean, been here for two years. Since arriving, she has not grown. Maybe now she is five years old.'

We guess her weight to be less than 150 pounds, and she is too small for a normal five-year-old Asiatic black bear. But how heavy should a five-year-old Asiatic black bear be? There is no literature here in Kathmandu. The next day we depart for the USA. As we fly over the Baluchistan Desert, I look down on the Afghanistan desert where our Yeti Volkswagen had burnt out fourteen years before.

Two days later, I am pulling open the brass handles of the doors at the Smithsonian Institution's National Museum of National History, doors polished by thousands of tourists' hands. Childhood dreams had me bringing something to this repository. Now accompanied by Jesse and Jennifer, we cross the ropes that funnel visitors towards the dinosaurs and head to the collections. The curator of mammals, Richard Thorington, is out.

Bill, his assistant, leads through the labyrinthine halls with case after case of specimen boxes, giant white blocks stacked up against the walls. In these, on sliding trays rest the mammals of the world, waiting for their appointed moments for their existences to be made into knowledge by the furless mammal whose another defining feature besides putting other mammals in boxes is upright bipedal walking. It is hard to look like a scientist while carrying my specimen in a red nylon stuff sack and accompanied by a two-year-old kicking a ball of paper along the floor. What's wrong with coming to look at specimens with your kid? If the skull I carry is added to this collection, it'll be as much his contribution as mine.

ᖇ

MUSEUMS HOLD THE PAST FOR PRESENT AND FUTURE USE. From life now gone they gather artefacts of aesthetics and the heritage of miscellany. As time moves on, museums allow us to go back in time bringing different times together.

Museums began as temples dedicated to the muses, places to inspire people, gathering the magic out of life experience. Ennigaldi-Nanna's Museum (in modern Iraq) was the first museum to gather artefacts for public display. She was the daughter of the last Babylonian king. Perhaps knowing that her heritage was on its way out, she dug up shards with her own hands that gave evidence of the wonder that Babylonia was, seeking to record a story that began with her ancestor Nebuchadnezzar. To allow the future to go back in time, she described her collection in three languages.

But earlier than creating places for things gone, people gathered the living. Zoos predate museums by millennia. Princess Ennigaldi's ancestor, King Nebuchadnezzar, built his hanging gardens of Babylon to house live wild animals and birds. Even 3,000 years before that, the Egyptians caged hippos, elephants, baboons, wild cats, and ungulates for display.

Across centuries, private gatherers with financial means or hard collecting work have assembled 'wonder rooms'. Many of these collections upon their gatherer's passing were given to the public, and the concept of one place to which would come items from around the world and back in time became more common. The British Museum in its founding purpose was to study such artefacts, gathering 'the interesting things' from their empire. Should anyone other than a scholar want to enter that museum, she or he first had to request their permit by letter. And again in England the gathering of the living had begun before the gathering of the gone, since a century before the British Museum, King Henry the First collected camels, lions, and leopards to create England's first zoo.

As the general public began visiting museums, at times the line of propriety was crossed. In a city that advances itself to be civilized, a display in 2005 at the London Zoo had people wearing fig leaves. In 1906, the New York Bronx Zoo displayed a Congolese pygmy— a live member of our own species—as 'the missing link'. Ota Benga had been 'collected' in the Congo Free State as a 'specimen'. The *New York Times* reported Benga was viewed in a cage that also held a chimpanzee and a parrot. On 16 September 1906 alone, 40,000 people gawked at him. New York's black clergy protested, 'Our race, we think, is depressed enough, without exhibiting one of us with the apes.' But the *New York Times* editorialized, 'As for Benga himself, he is probably enjoying himself as well as he could anywhere in his country, and it is absurd to make moan over the imagined humiliation and degradation he is suffering.'[1]

As I went on my footprint expeditions, one question I've wondered over the years is: what would be done with a Yeti if I found it? I might try directing the action, but certainly I would lose control. Would my Yeti be welcomed as a member of society, a hominoid, or put into a zoo as an animal? Or, if only one individual was ever found, would it be pickled and put into a museum? Though I carried on with my search, I often asked myself whether I really wanted to find a live Yeti.

∽

BILL STANDS BEFORE THE DRAWERS OF OUR HYPOTHESIZED SNOWMAN. The skulls Bill takes out are meticulously washed, each penned with a number in black ink on bones sanitized white. These bears don't have to fight for territory; the human species has assigned to each a little box in its drawer in a bigger box in its Latin-named plywood-box jungle. I unwrap the newspaper padding from my red nylon bag and reveal a yellowed skull, dried muscle, and a few tendons attached to it.

While my left hand holds the mystery, my right picks up a skull whose identity is ascribed. I look into bony faces and big eye sockets. I almost expect sinews and breath to flesh these bones anew. Holes perforate the skulls, entry and exit channels where nerves and veins once ran carrying out their appointed tasks. My fingers explore the skeletal ridges, concave and convex undulations of calcified structure. How insulting to play with the head of a once-fierce bear. I look into our skull's gaping nose hole;, here they call it 'nostrum'. I doubt the bear had a name for this, its most important sense. What would it have thought (how do bears think?) when the message of peanut butter first came through this hole?

What features differentiate the beasts we call 'wild' from those we call 'not wild'? How does a taxonomist differentiate between a bear in the jungle and a bear in a zoo? (Not at all.) How do two bears define themselves in relation to each other in a zoo? (Human conditioning causes them to learn to live together.) The white plywood boxes may hold value, but in collecting and caging animals, do we put up fences that prevent us from understanding life? As we collect what these boxes hold, do we ask how it stacks up to our values—or to the values of those life species?

I am pulling out skulls, wondering to myself what to wonder about. Bill shuffles, saying, 'I know little about bears. My specialty is shrews. For each family and genus key features exist in the skulls, and so I don't know what's critical for bears.'

Of course he doesn't know, I tell myself. The last time anyone walked into the Smithsonian with a skull from a bear previously unknown was a century before; it was when the Kodiak bear was 'discovered'. Jennifer and Jesse patiently wait while I probe trays. Some skulls have a hole behind the eye socket; others have what seems to be the same hole in another place, and one skull doesn't have an equivalent hole. Did that hole-less individual have trouble using some part of its mind? Dramatic differences exist among ridges down the crest of the skulls. Some are smooth while others have a ridge that sticks up half an inch or so; one has a ridge and a half. Sherpas say that the Yeti has a pointed head. Could that be an extreme ridge crest?

Bill is pulling out trays containing the skulls of other Himalayan bears; sloth bear (*Melursus ursinus*) skulls and Himalayan red bear (*Ursus arctos isabellinus*) skulls. Their skulls are broader than ours with smaller nose holes and bigger eyes. The only close fit to our skull is the Asiatic black bear. So our skull's genus is certainly *Selanarctos*, and these white boxes attest to only one species in this genus: *thibetanus*.

Bill suggests that teeth are probably critical. Our skull has the *Selanarctos* configuration of molars, premolars, canines, and incisors. The teeth in our skull are worn, so our bear was mature. But our skull is about 20 per cent smaller than any here in the Smithsonian and its molars are also smaller. This supports a tree-bear hypothesis: a 20 per cent smaller skull for a bear of the same age might correlate to a bear half the physical size. The teeth size of our bear equals *Selanarctos* specimens from Iran and Japan found on the peripheries of the species range. Our bear must be an adult *Selanarctos*, but why is it so small? Is it nutrition? Does overpopulation of bears in the Barun produce stunted bears? Or, is our bear an abnormal individual? Or, is there a genetic difference? No skull in this collection comes from Nepal, so there's no possibility of a direct comparison.

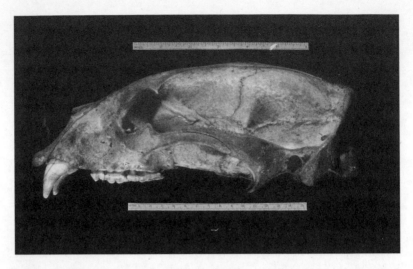

8.2   Our Skull of the Rukh Balu (*Selanarctos thibetanus*)
Source: Michael Meador

The next day in the Smithsonian mammal library I find little in the literature except for the fact that the species' 5,000-mile range, extending from Iran to Japan, allows it to live in warm India as well as in cold Russia. On one sheet of paper I list all the references. In five hours I have read all, except for four documents that are not in the Smithsonian library.

The literature says adult bears can weigh from 200 to 400 pounds. But weights even at 200 would be greater than what Lendoop says need 'two men to carry'. According to our guess, the five-year-old in the zoo weighs 150 pounds. That is 25 per cent less than the weight of *Selanarctos* from Japan, Taiwan, and Iran. The 200- to 400-pound weights fit with ground bears for which five or six men are needed. But the literature also says that for bears, generally weights have little significance. Mature grizzlies can weigh anywhere between 300 to 1,000 pounds. Small size only indicates poor nutrition in the first year of life.

Then another puzzle emerged. The literature reports that *Selanarctos* is aggressive across its range. (I know that, for Uncle John once had one of these bears lift off a colleague's face which then, as this fellow lay bleeding on the ground in the jungle, John had to stitch back on with his sewing kit.) But the tree bear is reportedly not aggressive. Can the tree bear be *Selanarctos*, as suggested by the skull, yet not possessing this characteristic the genus always seems to have?

Then in a footnote in R.I. Pocock's *Fauna of British India*, there is casual mention of a tree-living Asiatic black bear, *Selanarctos arboreus*. It describes a specimen submitted by Oldham. J.E. Gray, Pocock's predecessor as the curator of mammals at the British Museum, categorized this bear by examining one specimen that had been sent to the museum in 1860 by Oldham, a naturalist working out of Darjeeling. A small bear living in trees, a bear whose name is tree bear! In this footnote, Pocock outlines his thoughts. In 1860, Gray had also been puzzled by a bear that lived in trees, *Selanarctos arboreus*.

Eighty-nine years after Oldham presented his skull to the British Museum, Pocock compared it to the other museum skulls and found nothing to warrant its classification as a separate species or subspecies. But in those intervening years Oldham's field notes and the bear's skin, which was also given by him, disappeared. Without the notes and the skin, Pocock had only Oldham's assertions, Gray's classification as *arboreus*, and one skull. That skull did not seem dramatically different, only smaller. So Pocock removed the *arboreus* species designation. But maybe Gray and Oldham were right and Pocock wrong?

The name *arboreus* means 'tree'. Did Oldham see the bear in trees? Was he translating 'rukh balu'? Darjeeling is sixty-five air miles from the Barun Valley with similar altitude and habitat. One hundred years ago an unbroken jungle and this similar dense-forest habitat would have extended from Darjeeling to the Barun.

Might Gray's article be here at the Smithsonian? Good libraries are wonderful; an obscure datum sits, dusted regularly, waiting to be read. The skull waited for seventy-two years in England. Pocock looks at it. It waits for another forty-one years on a page in a journal. And one place where that single page might wait is in this room, for 114 years. Data sits in libraries around the world waiting for a century (maybe a thousand years on a pottery shard) to build knowledge, to be used to advance our seek-to-know-it-all species.

High on a grey metal shelf I find the annals of the *Royal Zoological Society*, bound in black leather and embossed in gold print; there were more than a hundred such volumes. I pull down the 1869 volume. And yes, before me is independent evidence that a field researcher gathered. He reported of a bear in trees and sent a specimen to his museum—1869 is two decades before any published Yeti fancies.

My eyes now race across the yellow pages. There is hardly any discussion of *arboreus* on the page; all I find is a footnote. Oldham found a small bear that he thought was different. How small? Oldham does not say. Why did he think it to be different? Oldham does not say. He collected the skull and skin and sent these to England. Gray concluded that its different size and coat were important. But Pocock did not agree; he found no unusual features when he focused just on the skull. The annals of the *Royal Zoological Society* do not say why Oldham and Gray called this bear *arboreus*. The possibly helpful notes as well as the skin are gone.

But my fieldwork gives a connection: a report, from roughly the same region of the Himalaya, of a bear whose behaviour is life in the trees. Behaviour in trees. Much that is distinctive about the tree bear is behavioural. Villagers do not claim a skull difference; museum curators worry about that. Villagers do not suggest a different genus or species. What affects village life are behaviours—like wiping out a crop in a night. It is behaviour that makes footprints.

Sitting in an armchair by the grey water cooler, I go through *Selanarctos* articles again. The librarian comes for the second time to tell me that the library is not about to close, as she had said earlier, but is now closed. I keep writing. The majority of information comes from the northern Himalaya. And in reports from Russia, I find a note that when walking in snow *Selanarctos* sometimes uses submerged branches to support itself. I also find that compared to other bears this genus places hind paw into forepaw tracks more consistently. From farther east in the Himalaya are reports of *Selanarctos* as an agile and delicate climber.

As I look down from the window to the traffic on Constitution Avenue, new facts and old ideas move in and out like traffic. But the person standing at the end of the corridor tells me that in a few minutes the security guards will be at her side. Near the Washington mall, hot-dog trucks are pulling down their aluminium side walls and away from the curb. For my family that has been away from the gentle hills of home for two months, it is time to drive back.

ৎ৶

4 APRIL 1983. THE POSTMAN LEAVES TWENTY-SEVEN BOXES OF SLIDES from Kodak in our big red mailbox. Might flecks of sand have gotten into my camera at the beginning of the expedition and on the film rollers scratching every roll? Will the rolls be black because of a flawed light meter? All day the boxes sit on my desk, one calamity tumbling on another in my mind. That night, after Jennifer and Jesse go to bed, I put new logs on the fire and brew hot chocolate.

It is more than simply looking at pictures—I want to relive the journey. In today's age of digital photography where you can instantly see what you have shot, it is hard to revert to the earlier mindset of not knowing about one's photographs. But earlier, when the shutter was pressed what was captured was not known.

In 1949, when Dad went on that first expedition permitted into central Nepal, he took only four rolls of film in three months (he bought all the rolls of colour Kodak film possessed in inventory by that photo store in Delhi, India film was so rare), and he had to wait four months after the expedition before he could see his images.

Slides in each box that I am about to open number one to thirty-seven. In selecting a box I do not know whether it is the first of the journey, nineteenth, or twenty-seventh. An expedition that unfolded day-by-day, will reassemble tonight backward and forward. I start projecting the slides; thankfully, the pictures are sharp with good exposure. In box thirteen, in the middle of that roll, the footprints climb out of high bamboo. The next slide shows delicate but perceptible disruptions revealed of moss on the cliff rock.

The next slide: the thumb imprint! The next slide again shows the footprint, the thumb once more distinct from the four digits. Two months ago, those distinct digits were Yeti-proof. Only primates have thumbs. I start over with that box watching the slides cross my screen. Then, as I watch, I see nail marks. Two months earlier—indeed minutes earlier with this same slide—I did not see these. I swore to Bob there were no nail marks. Of course, that day my brain was starved for oxygen, body fatigued from climbing through deep snow. Having proposed that condition for others, why didn't I see this then?

Even now I start questioning what I see before me. Might I find some slide that shows no nail marks? I stop what I am doing. If I was just now, might Shipton have also selected (though unconsciously) the print with no nails? New ideas collide with old, hypotheses of the tree bear with the Yeti. On the screen, though, footprints tell their tale. What is clear is that my foot has nails. What is also clear is that the other photograph taken by Shipton (of Michael Ward looking at the trail of prints) does not indicate how representative was the one photograph that launched the global Yeti quest.

I return to my slides, seeing now the feature to be looked for. Does this fit my other evidence? Bringing from the attic the paws purchased from Lhakpa, I pull out the small green Bic lighter I had laid in the track to indicate the size when taking the photograph. I move the projector to adjust the image on the screen to the size of the lighter. When 4 feet away, the lighter on the screen equals the instrument I hold.

The left forepaw then fits precisely against the picture. The print in snow is identical to the paw in my hand, nails on the paw matching indents in the snow. The only difference is the snow print is longer. Looking at the image on the screen, I see a faint line two-thirds of the way down the print in the snow. I pick up the other paw, a hind paw. That fits the back end of the print on the screen. Holding the forepaw and the hindpaw together, I observe that they perfectly overlap what is on the screen.

I did not see a tree bear make the print in the snow, but with the dried paws against the image on the screen, there is no doubt left about the maker of the snow tracks. On the screen is what I identify with passion as a Yeti. In my hands are two tree-bear paws.

Will Shipton's as well as Cronin and McNeely's prints also fit this way of setting my dried paws over their images in the snow? Photographs of their prints are at my office a mile away. I jump into my old VW van and bounce along the dirt road to the office. On the way back, a flat tyre cools my excitement. The night shines as I step out of the van. Our house is half-a-mile away; the spinning wind generator making our electricity sends a gentle 'thwup, thwup, thwup' across the mountaintop meadows. As I change the tyre, spruce trees whisper and the brittle grass of last year rustles with the turns of rusty lug nuts.

Almost an hour after it first walked across my screen, the bear track is alight again. I compare it with Shipton's photograph. The photograph and dried paws do not immediately match as there are no obvious nail marks. Also, the ordering of toes is not uniform with the thumb placed differently. But as I look more intently at the centre rather than the captivating front, with dried paws over the picture what I see changes. There might be nail marks in the middle. One deep niche is apparent on the left side, and under the rightmost nail of the paw I hold seems to be another mark in the snow. These fit the paw in my hand—and if I slide the dry hind paw further back, the length also approximates closely. The long-standing enigmatic Shipton print has a provable bear explanation if the requirement of front nail marks are taken away, and that might happen if the bear's weight had not pressed down so hard that these front nails would not show. Two of the rear paw nails do show.

Then I hold up Cronin's photograph, taken a few miles from where our prints were found. Sizes correspond: our track is 6.7 inches long and 4.8 inches wide; theirs 9.0 inches long and 4.8 inches wide. To explain the differing lengths, as with the Shipton print, again the math works by moving the hind paw back, almost exactly as much as with our print. The thumb-like inside digit on theirs and ours is in the same place. For them, the snow was soft and so no evidence shows in their picture of nails. I would very much like to see their plaster cast, the one Bob once held.

With new logs on the fire, I settle into the sofa—the slide still shining on the screen—and again read their description:

> Shortly before dawn the next morning, Howard climbed out of our tent. Immediately, he called excitedly. There, beside the trail we had made to our tents, was a new set of footprints. While we were sleeping, a creature had approached our camp and walked directly between our tents. The Sherpas identified the tracks, without question, as yeti prints. We, without question, were stunned.[2]

That is all they give. Why is there no discussion of the evidence and its positives and negatives? They had a Yeti print and, after presenting just the fact of it, they, like the animal itself, walked away from it and the questions it raised in the narrative. The answer, of course, is that scientists, especially rising ones, are wary of a serious discussion of the Yeti. Cronin and McNeely are scientists. McNeely now is a highly esteemed one headquartered at the International Union for the Conservation of Nature, based in Geneva.

☙

NICK RETURNS FROM NEPAL A WEEK LATER. From the tone of his voice on the phone call it's clear he has more to share than memories of a month in the jungle. He knows nothing of the Smithsonian skulls, of *arboreus*, or the slides. Minutes after entering the house, he's talking.

'The prints I found in the Langtang Valley were like those we saw on the ridge—thumb-like marks too. In Tarkegaon, I found the skin of a black bear about the size of the zoo bear. The man who owned the skin had killed the bear, so I questioned him. On his own, he used the words 'rukh balu'. Yes, rukh balu. The surprising thing is villagers also claim the presence of a large bear, *bhui balu*! Another villager showed me the skin of a ground bear. I took pictures of everything.'

Do two bears exist not just in the Barun but across the eastern Himalaya? In how many valleys? Is Nick's Nepali good enough for his reports to be trustworthy? Two months later, a letter arrives from Bob: 'Since we were together in February, on three treks I picked up village reports of tree and ground bears: the southern slopes of Himalchuli, Rowaling Valley, and Sikkim.' As I had been trying to put the pieces together, I've written to Cronin twice asking if he and McNeely had found two bears, but he hasn't replied, and I've called his phone. I've also sent letters to McNeely in Switzerland, but there is no answer from him too—did they find reports about tree and ground bears?

Working from field notes, I pull together the known fragments.

1. Reports of the tree bear come now from the Barun and four regions (Fleming found three of these). These reports cut across ethnic groups that have little contact with each other; yet, despite their cultural differences, tree- and ground-bear descriptions are consistent.

2. A small bear lives (and has lived for two years in the Kathmandu zoo) and has not grown. It fits the description of tree bears. Of particular interest about this zoo bear is a dropped inner digit on its front paws.

3. We possess an apparently mature skull and two paws from another such small bear; this bear being Lhakpa's Barun bear.

4. In 1869, Oldham reported the existence of a similar bear in Darjeeling, sixty-five air miles from the Barun Valley, and called it *arboreus*.

5. Nick photographed the skin of another such tree bear in the Helambu Valley and saw the skin of a larger bear alleged to be a ground bear.

6. The photographs we took in the Barun indicate that this tree bear makes tracks that are primate-like. It seems to have a hallux and the ability to walk bipedally, thus fitting the Yeti stories—yet we know the prints have been made by a bear.

7. Nail marks and hair suggest this bear (the hair sample collected from the nest matches *Selanarctos* hair in the Smithsonian) is dexterous at climbing trees and makes nests breaking bamboo one-inch in diameter at uniform heights. It also seems to make nests in oaks, and this suggests superb tree-climbing ability.

8. Some of the aforementioned traits are reported for *Selanarctos* but not all. But it needs be noted that *Selanarctos* has never been described scientifically. Most *Selanarctos* evidence has come from hunters before World War II. The Russian evidence, not yet studied, appears to be more scientific.

Is the tree bear a subspecies of *Selanarctos?* Are there genetic differences not visible in the skulls? Are tree and ground bears the same? Might tree bears be female and ground bears male? Or, could the tree bears be juveniles and ground bears adults? Also, could tree bears be runts and ground bears normal?

What is particularly puzzling about the Cronin or McNeely evidence is their descriptions of aggressive bear behaviour (Cronin spent half-an-hour telling me on the phone and told the same to Fleming). Those reports indicate ground bears. How could Cronin or McNeely spend two years in these jungles and never discover nests? That they missed the tree bear may not be that surprising; I missed it too, and we found it only when we knew what to ask. It seems incredible that a bear should remain scientifically undetected. But maybe it is not. The era when scientists were looking for new animals ended in the mid-twentieth century; Nepal was closed then.

ᐧᐧᐧ

AS A ZOO GATHERS DIVERSITY OF THE ALIVE and a museum of the dead, mountains gather a diversity of nature. Across the earth's flat lands habitats change incrementally, getting colder as distance increases from the equator. But when compared to the slow ecosystem change from the curve of the earth, temperature change in mountains is rapid. Ascend 1,000 feet, and the temperature typically drops by three degrees Fahrenheit (depending on the humidity).

As a young boy I noticed that flowers bloom at different times over the span of the 1,000-foot climb from Woodstock School to our home at the top of the hill. Then while in college in America I realized I did not need to go to New England to see vivid autumn colours; I could drive to the nearby Appalachian Mountains. I began to see that an aspect was missing in my biology textbooks. The textbooks said that the tropics was where life was most robust;

a diversity of species created that strength—but it was clear to me that life was robust also but in a different way on the mountains as I climbed higher and higher.

Life at the bottom of mountains was robust in terms of the wealth of species found there, but life was also robust at mountaintops where fewer species lived. Life forms up high had a strong ability to endure harsh and always varying conditions. Both types of robustness were strengths, but the strengths were not the same. The wealth of species was what gave strength at the base of mountains with genetic diversity; diversity across many species gave strength. Though only a few species had the ability to deal with the fluctuating climatic diversity of mountain heights, those few were hardy. (Gentians that can burst through snow and still bloom show strength—most flowers lose their ability to bloom after freezing has begun.)

Moreover, temperature is not the only aspect that keeps changing in mountains; life forms higher up have to also adapt with regard to food, moisture, oxygen, and soil. Turning a ridge, the sun's energy on a slope can decrease, maybe by as much as half. Change in moisture can be similar, as ridges dramatically change rainfall levels. Sun, moisture, oxygen, and soil, even the stability of the slope are in flux in mountains. Yet life not just survives, but thrives. Finding no terminology in the biological literature to explain this strength, I offer a new term: bioresilience, a term I introduced in 1995.[3]

Bioresilience is the ability of a species to adjust to change (whether related to temperature, moisture, sun's energy, or soil) in the course of its life cycle. Species whose individuals can withstand great change have greater bioresilience. (Some species survive by mutating to new species for the new niche; other species have already mutated into hardiness and so they can survive a wide variety of life circumstances.) With biodiversity, it is the number of species, the genetic diversity, that creates a net with many strands back and forth giving strength to the life net.

With bioresilience, however, a different strength is emphasized, which lies not in the number of strands but in the life strength of that particular genetic strand—the difference being between a net that is strong because of many fragile strands and a rope whose strength is in one strand.

Biodiversity has many species, each species filling one biological niche. Bioresilience has one species filling multiple niches, exemplifying its ability to cross econiches. Biodiversity's strength comes from genetic complexity, multiple groups of DNA. Bioresilience's strength comes from the ability of individual species with one DNA to accommodate changes in living conditions.

The climate needs to be stable to generate biodiversity, whereas the climate and other features fluctuate to promote bioresilience. Consider the adaptability life must have in mountains where temperatures can flip from freezing to near-boiling in one hour. (At higher elevations not only is there a more rapid build-up of heat from the sun passing through a thinner blanket of air, but boiling points are also lower due to decreased air pressure.) In the tropical world one season is much the same as months that came before and those following; due to this stability, life delicately evolves and fits tightly into niches.

Rising altitudes have a thinning blanket of air to buffer temperature change. So when the sun goes down (because plants cannot run indoors for shelter), in this fast temperature changing world not only must vascular systems empty fluids to prevent rupture from freezing, but in shutting down they must be poised to bounce back to maximize time when the sun's energy returns, driving metabolism, respiration, and photosynthesis. And, in these shortened periods, for the continuity of a species, reproduction must not be disrupted. The ability to accommodate change is essential when the climate fluctuates. This creates for life forms the feature of bioresilience.

In bioresilience, the focus is on function. Gender differences lose their decoration. Food sources, which may be deprived for half the year, must harbour nutrients from the harsh world in roots and rhizomes, body fat, or underground caches (for example, to accommodate such needs some species minimize metabolism through reduced activity or hibernation). Bioresilient plants avoid the complex reproduction rituals found in the biodiverse tropics where some times plants prepare for reproduction a year before they procreate (for example, the high-altitude flower-bud primordia).

Robustness of each individual in bioresilience crosses a series of niches—this is achieved in biodiversity by multiple species where each species fills one niche. The diversity of species decreases as altitude is gained, but to accommodate this, the resilience of each species grows greater. Bioresilience is not better than biodiversity; the two strength aspects complement each other in life and suggesting one is better than the other overlooks how the complexity of life carries on in our living world. Without bioresilience, hardiness of life would be limited to a girdle around the planet.

Because we are a whole planet riding together, conservation management (that utilizes natural strengths to protect nature) will be strengthened by conserving bioresilient life in a complementing way as it has been focused on conserving the biodiverse life in recent decades. Valuing bioresilience may now be especially urgent. For in the world of the future, as climate system constancy is lost (somewhat analogous to the thinning of air as a mountain is climbed), hardiness of species might be the life aspect that will give the planet the ability to ride the macro change underway to regrow potential for all species, to position repositories of life that can sustain all life.

Now our planet irrefutably spins into an era of unpredictable eco-change. To keep our understanding of life (biology) current, the concepts we use to guide our response can evolve from what we have been using thus far. Conservation can hold a dual view—it is not an either/or between biodiversity and bioresilience, but from both may come a depth of understanding for life to carry on. This must not be viewed as a battle where the frontline is biodiversity protection—biology can expand to an understanding of creating flexing systems. While focus must certainly be on preparing pockets of protection (especially delicate life forms which are natural treasures), preservation will be wise to also include life with all the adaptive capacity it can muster.

Like the bounce of any ball, this world will not rebound from the forthcoming changes unless we engage the flexibility of our whole sphere. One way is with one species adapted to each niche—biodiversity. Another way is to have species that are able to cross multiple niches—bioresilience—fitting many conditions. We homo sapiens step into a new habitat in this new age we have created: the Anthropocene. The species we have often viewed as invasive may be links to bind together our world such that it is persuasive. And so, God bless those species that are everywhere: cockroaches, crows, zebra mussels, and yes, humans too.

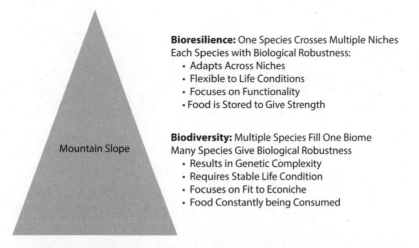

Bioresilience: One Species Crosses Multiple Niches
Each Species with Biological Robustness:
- Adapts Across Niches
- Flexible to Life Conditions
- Focuses on Functionality
- Food is Stored to Give Strength

Biodiversity: Multiple Species Fill One Biome
Many Species Give Biological Robustness
- Results in Genetic Complexity
- Requires Stable Life Condition
- Focuses on Fit to Econiche
- Food Constantly being Consumed

Mountain Slope

8.3   Bioresilience and Biodiversity: Comparison of Attributes

Source: Author

The challenge then becomes about how to carry out this new vision. Nature preservation principles and practices must change. We must become resilient now in our management approaches. Conservation action is thus not a search for a lost wild man; rather, it leads us to the search for a new way for people to engage with the wild.

Notes

1. See   http://www.nytimes.com/2006/08/06/nyregion/thecity/06zoo. html?pagewanted=all&_r=0.
2. Edward W. Cronin, *The Arun: A Natural History of the World's Deepest Valley* (Boston: Houghton Mifflin, 1979), p. 157.
3. See Daniel C. Taylor, *Something Hidden behind the Ranges* (San Francisco, CA: Mercury House, 1995), pp. 137–8, 201–4, 213–16, 221–5, 273. In this earlier book, I first outlined many of the concepts developed more contemporaneously in the present volume.

nine

Evidence Slipping Away

December 1983. I had walked through these same glass doors in the summer of 1969, the C Street entrance of the US State Department, my first day with the US Foreign Service. I had been assigned to work on the 'population problem' in the Himalaya. In the fifteen years since, two billion more people ride the earth, a 40 per cent heavier load, adding a burden in such a short period greater than the planet had ever experienced, with each person believing he or she will gain a better life. World gross domestic product tripled from USD 800 per person to USD 2,600. *Homo consumpticus*[1] pushed wilderness to near extinction worldwide.

Glass doors close behind me, they still have the delicate black borders that look haunting like lines that rim documents of bereavement. The security guard directs me to the elevators for those of us who are attending a state luncheon with the US vice president and the king of Nepal. Stepping to a corner to gather myself, an empty pocket tells me that my comb must have popped out just now when I was paying for the taxi. I'm sweating from chasing down that cab after an interview at the National Public

Radio. Fortunately, in the interview Noah Adams didn't ask, 'Is your new bear the Yeti?' I am holding the suitcase with the bear's skull. Where can I put it? I wonder.

As I look for some place, through the glass door comes Lila Bishop, whose husband Barry was part of Hillary's 1961 Yeti-hunting expedition; and in more recent years I've shared memorable climbs with him. 'Dan'l, why are you lugging a suitcase to a vice-presidential lunch?'

'Lila, where is yours? Mine holds one bear skull and two dried bear feet. Did you bring something for the king and the vice president … maybe yak cheese?'

Grabbing me by the elbow, she leads me to the elevator, where we stand at the back while I use her comb, and she straightens my bow tie. As the elevator doors open, Secretary of State George P. Shultz steps out of another door, followed by two aides. After some confusion at the name-tag table, the receptionist agrees, with a question about security with regard to bear paws at a vice-presidential reception, allowing me to stash the case behind her table. A harp plays as we step into another world, a string quartet accompanies, and gilded pictures embellish the walls.

Barry Bishop, the vice chair of the Committee on Research and Exploration at the National Geographic Society, comes over. 'Congratulations. I read two days ago about your bear discovery in the *New York Times*.'

With pleasure, Lila tells him about the paws and skull behind the desk. Barry spots Russell Train, the president of the World Wildlife Fund (WWF). Train asks me to make an appointment to discuss the discovery. Meanwhile Barry is pressing me about my plans to follow this up. He reminds me that I'm not a biologist, that folks are going to connect my bear with the Yeti, and suggests that I must get a bear scientist.

Since I don't want anyone at the National Geographic to link my specimens with the Yeti until I have proof, I push back. I tell him that with Bob the scientific aspect is under control—plus,

Selenarctos thibetanus has never been scientifically studied, so there is no suitable bear expert.

The reception line moves closer to the vice president and His Majesty. The king quickly figures out what Barry and I were debating and wishes me good luck. The vice president nods quizzically. Barry continues to press the point that without a bear expert we will not be able to raise money as well as that article in the *New York Times* and the National Public Radio prove nothing.

I try to assure Barry that we shall collect both types of skulls, and to do that I'll walk back and purchase both. Barry points out that nothing is so simple, and he is right. He cares, and I need his help. He and I have shared many uncomfortable nights in tents from the Arctic to the Himalaya. Barry offers to introduce me to John Craighead, noting that if either John or Frank Craighead participates, bear experts around the world will pay attention.

ඉ෮

BACK IN WASHINGTON TWO MONTHS LATER, I walk into the WWF office. Russ Train welcomes me and introduces Curtis Freeze, a zoologist working on their Amazon project.

I explain what we found east of Everest and how Nepali villagers claim to have seen two types of bears: a small tree bear and a larger ground bear. The ground bear seems to be the known Asiatic black bear. But the tree-bear descriptions do not fit. It's easy to discount these as stories, except that, as I open my little green case, I explain how the skull is uniformly smaller than *Selenarctos* skulls at the Smithsonian and New York's American Museum.

As Train and Freeze inspect the skull, I explain how other reports from five places in Nepal also talk about tree and ground bears, adding the evidence of footprints and nests, evidence that complicates what is known about this *Selenarctos thibetanus*.

Freeze presses back, as in a legal cross-examination, pointing out that what we really have are stories, and stories do not prove hypotheses—indeed by my own admission our skull is not significantly different.

I explain that I'm not suggesting two species; there is no conclusion about that yet. But discrepancies such as my smaller skull require study, and to do good research I need money. Yes, there may be an opportunity to differentiate subspecies, but there are facts about unusual behaviours regardless of taxonomical findings.

Freeze presses further, asking about my qualifications, about whether I had ever previously worked with bears. I again explain my ideas of behaviour, of connecting different village reports. He then puts it bluntly: what I have raised are questions, and to answer these scientists who have worked on these questions are needed. As the meeting concludes, Train suggests that I send a funding proposal, reminding me that the WWF focuses on preserving endangered animals, not finding new ones.

With an hour still before my appointment at the Smithsonian, I sort out my thoughts over a too-expensive hot fudge sundae. I've now knocked on the doors of several major conservation organizations. Not one views the questions I'm asking as important. And because my skull matches *Selenarctos*, I lack a hypothesis with evidentiary support.

I stir the ice cream and fudge. Why are professionals not interested in this? These confident believers remind me of some missionaries I knew as a child, people who defend a faith but are not open to exploring in faith. Action about the environment now focuses on priorities in the courts, boycotts, or scientific papers, and do not enter the great mysteries of fieldwork. Fieldwork begins with untidy postulates that are hard to justify. (The word 'Nature' originates from a Latin word, *natura*, which means 'conditions of birth'.)

That is Nature: untidy, the process of life coming forth. Thoreau was wrong—in wildness is not the preservation of the world for that omits the essential of preservation being in the way we live.

And *that*, the way we live, is truly messy. The wild of Thoreau is no longer with us. Our behaviours have changed the wild—grasping that new positioning is understood by more than merely science, and what produces loss of the wild is separation from it. Instead of Thoreau's end of living with Nature as a means to its preservation, wildness now is to be looked at rather than lived within, a separated world that gets joined when the lighting is perfect on TV, edited to show as often as possible males fighting, females nursing, or the two having sex. Through separation, wildness has been reduced to voyeurism, marked by a failure to remember that Nature is the process of life coming forth with ourselves in its womb.

I stir the chocolate and melting ice cream. Seeking to preserve wildness, we lock it off—though now it is actually impossible to lock off as people are everywhere. Wildness will be newly found again in recognizing that it is everywhere—as it always has been. It is in the process of going to it and learning from living it, as Thoreau did. Wildness is not on the fringes (a remnant valley in the extreme Himalaya) but in the openness of ourselves to the world ... which perhaps was what Thoreau was also saying.

I am sitting by the Potomac River, and even here wildness can be engaged by people who stop at this river's edge and step into life a little more, perhaps seeing a fish dart, bringing wildness into their lives, a world they don't control. The view that to enter the wild first requires wrapping in Patagonia and North Face clothing, and then getting on an aeroplane, this is really a going away, separation. Distance has been introduced, not very different from entering the wild by viewing through a glass screen, entering experience but under human control. Authentic wildness is found in choosing to walk home in the rain, embracing the absence of control and letting it soak into you. Wildness welcomes us whenever and wherever we want it.

I hurry across the Washington Mall, reminding myself that I am headed to the curator of mammals at the esteemed Smithsonian. I must let go of the ice cream and fudge in my thinking. Focus on the literature I've been finding. For example, no longer is my bear *Selenarctos thibetanus*. Recent DNA analysis placed the Asiatic black bear in the genus *Ursus*, where also live the grizzly bears, the North American black bear, and the polar bear. My bear is now *Ursus thibetanus*. Equally, the larger *Ursidae* family (that includes sun bears, sloth bears, spectacled bears, even giant pandas) is no longer related to the raccoon family. Thus, within *Ursidae*, my *Ursus thibetanus* is genetically closer to *Ursus minimus*,[2] the proto bear of five million years ago. (Did the proto bear exist at the time when *Homo sapiens* was coming forth?) *Ursus minimus* was distinctive due to its small brain and big mouth, features that today characterize its most direct descendant, *Ursus thibetanus*, a bear by comparison to other modern bears with an elongated and strong jaw.[3]

The big jaw muscle allows this oldest of bears to eat virtually everything—not only carrion, but when it can catch these alive: monkeys, serows, ghorals, wild boar, and even domestic cows. Invading farmers' fields is something that the nose (and also because it has little brain) cannot resist, especially when maize is ripening. Drawn in by that nose, insects and larvae are also dug out of rotting logs and from the ground, and there is always honey from bee nests (Pooh reminds us). It eats fruits of all sorts, developing in some regions proficiency at tree climbing as it scrounges for nuts.[4,5] Anything once living, my bear will eat.

Reports out of China (where its common name is moon bear) point to the significant amount of time these bears spend in trees, a feature not mentioned in the literature for the Himalaya but supported by its physique—powerful forearms as compared to other *Ursus* species and relatively weak hindquarters.[6] These reports also mention that when in the trees this bear makes nest-like features. It is not clear whether these nests are for sleeping or simply platforms from which they eat.[7]

And now about that bipedal walking. From my childhood I've seen these bears trained to dance. Men with a bear leashed on a chain went village-to-village. Usually, they led sloth bears, but sometimes it was the Asiatic black bear too. One man had trained his bear such that it would run after him around a circle. This bear was not doing just a few steps on two feet, or bouncing up and down, or shifting from one foot to the other; the man had his Asiatic black bear actually running up to maybe 30 yards.

As I proceed across the Washington Mall and walk towards the Smithsonian, I realize that in my next round of reading I need to better understand the dynamics of plantigrade feet (giving a platform on which to balance). The plantigrade feature is what allows two-footed walking, where the walker comes down on the hind part of the flexing platform, then goes forward propelled by the hinge at the toes. Being plantigrade improves balance—it gives plasticity at the end of the leg rather than a rigid hoof or cushioned paw of a dog or a cat. It also shortens part of the leg and causes bears and humans to be less powerful runners. To allow this ambulatory dexterity, bones of the foot readjust the five digits plus the bones of the sole.

How flexible might those digits be in the case of a bear? For a bear that trained its digits through its first years climbing trees, how would those digits splay out when the animal is on the ground walking on snow? Questions drawing my attention, I realize, are anatomical and behavioural, not taxonomic, always going back to the footprint puzzle. Just because the prints *look* like their maker is walking on two feet does not mean it actually is. Two-footed walking bears are trained. People walk in the snow on two feet because that is all we know. Why would an animal walk on two feet in the snow when it could walk either way? Walking on four feet would distribute its weight and keep it from sinking in so deeply.

ᏩᎤ

AT THE SMITHSONIAN, I KNOCK ON THE HALF-OPEN DOOR. 'Ahh, Taylor, the fellow with the Himalayan bears.

I'm Dick Thorington.' A smallish gentleman rises, hand out-stretched, and says, 'Sorry, I wasn't in the two times you came before. Glad Bill could help. Let's check your skull.'

Thorington leads through the hallway maze. 'Here we are, *Selenarctos*. One day we'll rename the case properly and call it *Ursus*.' Out slide the wooden trays I've come to know, the product of more than a century of collection across 5,000 miles of Asia—bears from the Baluchistan Desert, northern Kashmir, southern Russia, central India, Burma, China, Japan, Taiwan, and not one from Nepal.

'Hmm, general bone configuration similar, occipital crests almost identical, nostrum remarkably the same.' Thorington holds our skull in his left hand and a large skull from Kashmir in his right. 'Dentition between the two skulls is identical; molars, pre-molars, incisors. Yes, it's all there.'

'Dr Thorington', I say, offering him a micrometer, 'measure the molars of our skull and then those of your central Himalayan skulls. The teeth of the smallest Himalayan skulls, even juveniles, will be larger in all dental measurements than our skull. Doesn't a consistent difference in the size of teeth indicate some sort of separation?'

Thorington takes the micrometer. 'You're right about the size difference. Intriguing though it may be, it doesn't prove anything. True, bears that are from the central range like yours would be expected to be equal to others in the Himalaya, such as in Kashmir or Assam. Instead, your teeth are like skulls from Taiwan or Iran. You say the habitat where you collected this skull is a pristine jungle. The logical explanation is that your population is nutritionally, not taxonomically, different. It is likely that there is something in or not in the diet that causes the population near Mount Everest to be not as large as one might expect.'

'But that argument does not explain villagers' claim of two bears, one of which is large, in the same habitat. The Barun Valley is the most pristine valley in the central Himalaya. There is plenty of food. The habitat ranges from subtropical to arctic, bears climb up and down to take their pick of the food depending on what is in season.'

As we talk Thorington keeps circling back to remind me that everything I am convinced of, and also what Bob and Nick heard, must be treated as hearsay by the curator of mammals for the Smithsonian Institution. A skull can be objectively compared— and the evidence in the skull I have does not appear mysterious.

What I need, he tells me, is *sympatric specimens*. To compare speciation, specimens that differ morphologically must be collected from the same habitat. Skulls are needed from the same habitat. If two animals show physical differences and interbreed, they are subspecies; if they show differences and cannot produce offspring, they are considered to differ as species.

'So there is no value to what the villagers say?' I query.

Again, his reply is, 'Villagers are not scientists. What might appear to be two bears to them could be taxonomically one species. Villagers express ideas that are scientifically implausible. In the Himalaya, an example of such would be the Yeti. You must be careful. With this bear you aren't saying you've found the Yeti, are you? Of course not. Villagers use language differently from the way scientists do.'

I wince, wanting to share our hypothesis. John R. Napier, a predecessor of Thorington at the Smithsonian, has famously debunked the Sasquatch in a book titled *Bigfoot: The Yeti and Sasquatch in Myth and Reality*. In that book, he writes the following about the Yeti: 'Something must have made the Shipton footprint. Like Mount Everest, it is there.'[8] Napier then advances a theory to explain the Yeti footprint, suggesting that these super human-size footprints that were indelibly there may have been two humans walking on top of each other's footprints, thereby making Shipton's enigmatic print.

As we talk, Thorington encourages further study, saying that there is 'something about this skull, something I can't pin down'. He notes that though its physical features are similar to *Selenarctos*'s, in its *feeling* it was more delicate. He concludes by saying, 'If you really think there is something to the village stories, you would be wise to go back into the field. Maybe the skull you have is from the known *Ursus*, and the new is yet to be collected. Villagers could have mislabelled a skull, especially if they wanted to sell you something.'

But I know that while skulls have become the taxonomical standard, evidence is growing that species and subspecies could have identical skulls but still differ on the basis of other distinguishing criteria. Could it be, I ask, that a parallel might exist in the case of bears as with Nepalese wren-babblers. Physically, the two species look alike with nearly identical skeletons and plumage, but despite similarities in appearance they are distinct species. One species has a different song, different breeding altitudes, all of which are visible in their behavioural traits. Though physically indistinguishable, wren-babblers turn out to be different species due to their differing DNA—genetic difference is then revealed in their behaviour. Might not two bears look the same and still be different?

Thorington notes that it is possible, but such an instance in mammals is yet to be seen. The only way to substantiate this would be to submit material from both animals for DNA analysis. If I want to make that argument, I need to start doing a DNA analysis of my bears. It is, of course, DNA analysis that has recently changed how bears are categorized in genus and species, a differentiation until now done primarily by skulls. And as he and I talk, reiterating how particularities of occipital crests only become notable when the skulls are placed on sliding trays in white wooden boxes, he graciously offers to write a letter suggesting the legitimacy of my hypothesis. 'Perhaps a letter from the Smithsonian Institution may help with your fundraising,' he says.

◌

OUR FAMILY REACHES HOME THE FOLLOWING MORNING. As Jennifer carries a sleeping Jesse upstairs, I go out to the car to bring in our bags. A deer snorts and crashes down the hill through the blackberry bushes. Washington is 200 miles away, a world feeling more distant than the Milky Way light years above. A breeze rustles the spruces down the slope by the spring through which the deer just took off. Looking skyward, I ask, is it possible to hold back oceanic unfathomables? A sliver of a moon hangs over the ridge called Spruce Bars.

Thorington has strong grounds for scepticism; it is his role as a curator. Yet he remains open to learning. If it is possible for him, why not others? Is knowledge from science constituted by achievements, or is it a method? Is science a process of future-probing, or is it examining citations sitting as factoids? I look to the stars. Ultimately what is being evolved is the truth, where what is verifiable by one person's measure can be replicated by another. Ideas speak out from beyond the possible—that can be unique to one person, but it turns into truth when replicated again and again. Our species' journey is to build one upon another.

Jesse wakes at 6:30. I carry him out of the bedroom to let Jennifer sleep and start putting away the groceries of last night. Where is that little green suitcase with the skull, pictures, and paws? I check; it's not out in the car. Then a splash of remembrance: last night, when Jesse was crying, we stopped to get food from the bags to quieten him and I took the case out and set it on the ground by the side of the highway. I must have forgotten to put the suitcase back in.

Jennifer hears me on the phone asking for the Virginia State Police. She takes Jesse from me and looks out of the window into the wall of fog outside.

'State Police, Woodstock, Virginia? Yes, I lost a suitcase in the Shenandoah Valley along Interstate 81 early this morning when I pulled our car over to attend to my crying toddler. Could you alert troopers to search the road? It looks like an overgrown green briefcase.

'What's in it? Well, this may surprise you, but there is one bear skull, two bear paws, and 300 photographic slides. All are rare specimens from possibly a new species of bear in the Himalaya mountains. ... Yeah, I know it's hard to believe someone would leave a case like that beside the highway. Oh, you have a little kid? She doesn't like to ride in her car seat either? Maybe you understand how my son's screaming got me distracted.'

As I hang up, Jennifer turns from the window. 'I can't believe it. After all we've gone through for that skull and slides, they now sit somewhere.' She starts offering ideas: 'Did you have our name on the case? Do you remember where we stopped? Did anything in the case identify us? Call newspapers and radio stations throughout the Shenandoah Valley.'

What's really in the case are family investments. I can't look at her, so I walk past the old bed in the living room which, for lack of funds, she has made up with pillows and bolsters to look like a sofa, and step into the morning fog, walking aimlessly over the 400 acres I know so well. Pacing the meadow north of Noah Warner's barn, a hawthorn tree rears out of the cloud like a bear. I am left with only stories now, without the evidence to give credibility to the stories. The news is out on the *New York Times* and National Public Radio. It would be one thing if it was just a bear, but if people start suggesting a maybe-Yeti now when the evidence has disappeared, it will be like Tombazi's Yeti slipping into the bushes as he struggled to put on his telephoto lens.

Do I race back to the Himalaya and get more skulls? Or, do I hide in the West Virginia mountains? I walk through the morning fog. When I enter our house, Jennifer has a fire crackling in the fireplace and a Mozart Horn Concerto is playing. I call the Woodstock barracks again.

'I'm following up on a report I made this morning regarding a green case lost along Interstate 81.... Oh, it was you I talked to? Any news?' She says troopers have searched the road from Toms Brook to New Market. I call all the service areas and plead with them to check their dumpsters in case someone picked up the suit-case and then threw it aside upon finding a bear skull, paws, and hundreds of slides.

Days, then weeks, pass. Mid-March moves to mid-April. I have called radio stations and newspapers. The address on the suitcase is of my parents in Baltimore, so I've gotten to know their mailman and UPS delivery man. For the past month someone has always been near our phone.

〇〇

AS NEWS OF MY BEAR REACHES THE MEDIA, questions about the sasquatch are raised. In radio and newspaper interviews discussing my loss I can sometimes limit questions to an unknown bear from the Himalaya. But with remarkable frequency questions link the bear skull to the American Yeti: Sasquatch or Bigfoot. (Sasquatch has books exploring its legitimacy; I summarize here one especially delicious sleuthing trail. The Wikipedia entry covers both sides comprehensively.[9])

As best as can be lifted from legend, Native American tribes had a belief in a man of the forests. The name 'Sasquatch' comes from a Canadian newspaperman in the 1920s who brought this name from the Halkomelem nation.[10] In America, Daniel Boone claimed killing a 10-foot tall man in the forests of Kentucky. But a modern believing public became convinced following a short film of a beast loping along a riverbank on 20 October 1967 in Bluff Creek, California. A string of discoveries followed throughout North America, particularly the Pacific Northwest.

The Patterson film was near-definitive evidence for four decades despite the book by the Smithsonian's John Napier. The film shows a 7-foot tall hairy hominoid (apparently female—it seems to have breasts) moving in front of fallen timbers. Footprints in the sand were extraordinarily long: 14 inches. Established academics, including Jeffrey Meldrum from Idaho State University,

Grover Krantz from Washington State University, and Bernard Heuvelmans (the 'Father of Cryptozoology'), gave credence-building interpretations to the film.

But thirty-seven years after that film, and two decades after I lost my suitcase, in 2004 the film's authenticity started to unravel. Among the cracks in the story, the investigative reporter Greg Long published *The Making of Bigfoot: The Inside Story*, bringing forward the history of Roger Patterson and his larger history of falsification.[11] Accompanying Patterson that day (20 October) was Bob Gimlin, who also witnessed the whole sequence, and across the years had corroborated what had supposedly happened. But Long found other people, and finally Bob Heironimus. The following quotations from Long's book summarize what Long found:

> Heironimus hunched forward, put his hands on the table, and then stared straight into my eyes. He had young, fleshy features, soft ruddy cheeks, and a fleshy jowl.... He cleared his throat, 'I'm here to tell you', he said in a soft, country voice, 'that I was the man in the Bigfoot suit'.[12]
>
> We dismounted and unloaded the sack. I had to kind of sit down and put it (the suit) on. It was stiff from about here (he gestured from his waist up). They kind of helped me up and put the top on. Roger said, 'Just walk over and stand there in that section, and when we get ready to film, you just start walkin'.[13]

Then Jeff Long speaks about the interview he did when he filmed Heironimus in which the latter mimicked what he had done years earlier.

> I centered him in the viewfinder (of the camera). 'OK, Bob! Get ready ... set ... go!' Bob walked with long, purposeful steps, knees bent, arms swinging in long vigorous arcs, shoulders slumped forward, face toward the ground. The puff sleeves of his jacket added mass to his upper body. A sudden chill rippled down my nape. I blurted out excitedly to Pat [Long's colleague], 'I think he wore the Bigfoot suit!'[14]

In a separate interview Long talks to Heironimus's mother who had stumbled on the mysterious Bigfoot suit hidden inside the car. 'Well, after he got back, the next morning I was goin' to put some boxes in the trunk to go pick up some apples, and when I opened the trunk, here was this black thing layin' there, and I stepped back like that … then I discovered it was just—just the suit', she smiled.[15]

Then came this discovery by Long who, on 26 November 2003, called up Phillip Morris, a costume-maker and magician, who said:

> When I saw the film on television. I knew within seconds that I was looking at my suit. I knew it! That suit was the style of my gorilla suit'.[16]
> 'Wow', I [Long] said. 'So, Patterson sent you a check?' Morris answered, 'No, a postal money order. He also sent money for shipping. So I took one of my gorilla suits and we shipped it to him. … Then, not long after he would have received the suit, I got a call from him. Patterson said, 'I can see the zipper in the back,' I told him, 'Just brush the fur down over the zipper'. Then Roger wanted to know how to make the arms longer.[17]

In the years since Long's book, a larger discussion of the film unfolded on the Internet. Parallel debunking came from Kal K. Korff who did a meticulous analysis, finding a series of anomalies; most importantly, inconsistencies between the feet on the animal when the movie was made and the footprints photographed separately that were shown as the ones made by the beast.[18]

The Patterson film had stood as the equivalent to the Yeti's footprints in the snow: made and never explained. As the film got debunked, this evidence lost its validity. For me, losing my suitcase similarly felt like having lost my credibility. For years I had stayed loyal to the footprint quest, trying to avoid crossing the line and believing that a real Yeti, though tempting to

9.2 A Yeti Tracking the Human Search

Source: Dan Piraro

believe in, might really exist. And finally, when I had an answer in hand for what might make those footprints, that answer slipped away just when the evidence was beginning to be presented for analysis.

ᘒ

FROM THE NEPALI AMBASSADOR IN WASHINGTON comes a phone call. The Smithsonian is hosting a reception at the

National Zoo to kick off fundraising for Nepal's King Mahendra Trust for Nature Conservation. His Royal Highness, Prince Gyanendra Bir Bikram Shah Dev, the king's energetic brother, is the chief guest and there is a guest list that might have possible donors for me.

The National Zoo is a great place for a wildlife gathering where the pandas prowl from across the wine table. The WWF has done a superb job of organizing this event. Ed Gould, the curator of mammals, introduces me to Jack Ziedensticker, the zoologist who used to run their tiger project in Nepal. 'Jack, meet Daniel, the fellow who came up with the new information on *Ursus thibetanus*.'

Jack is dismissive, claiming there is no way a new *Ursus* species could be out there. The Himalaya has been gone over scientifically for a century. This is not some tiny, drab-brown bird with a small localized habitat. A bear couldn't hide undiscovered. Bears are not reclusive; they invade fields.

I try to defend, saying that in a way this isn't news—and that I am not suggesting a new species but new information. Villagers have talked of two bears for years. And now a skull is collected; additionally, nests and footprints have been observed and photographed. These describe bears in trees whereas previously we thought Himalayan bears were ground bears. These are differences needing study. And this contemporary evidence is supported by a citation in the 1869 *Proceedings of the Royal Geographical Society* that talks about bear skin and a skull Oldham collected in Darjeeling, 60 miles from our site. Oldham named his bear *Selenarctos arboreus*, indicating presumably that it lived in trees, a connection to reports today that talk of 'tree bears'.

Blessedly, our conversation ends as a friend, Mary Wagley, and her uncle, Charles Percy, arrive with warm hugs. As Jack walks away, I'm struck again by how hostile some people immediately become to this bear. I suspect that if the proposal was about a rodent, there would be openness. Like the earth's other large plantigrade animals, might bears walk too closely to humans? And though no one mentions it, might the thought lurking behind people's consciousness be that I'm talking about the Yeti?

But nobody knew or asked questions that made me talk about the lost suitcase.

෨෩

ON OUR MOUNTAINTOP, AS I WAS WASHING DISHES two weeks later, the phone rings. It's the newspaper editor in Woodstock. 'Dr Daniel Taylor? You remember we ran that second story last week on your lost bear artefacts? On my other phone now a caller is on hold; he says he has your suitcase. He won't give his name or phone number but wants to know the reward. Ooops, the light just went out for his line.'

The receiver in my hand is also suddenly dead, black cord curling into a black box that goes nowhere. My heart races. Where did I put the phone number of the Woodstock newspaper?

The phone rings again. 'Sorry, I was trying to salvage the other call. Yes, the guy seemed for real. He spoke simple English and seemed afraid. I asked him twice for his name. He wanted to know how much you'd pay. Then his phone went dead. I think he worried that when put 'on hold' I might be tracing his call. Maybe he'll call back. How much will you pay?'

'Pray he does. Tell him USD 250. I'll pay more, obviously. But I think a higher price might cause him to think there's something illegal among the stuff. Assure him this is legit—I won't ask his name. Whatever, please, please get that suitcase.'

'Let's wait.'

An hour later the editor calls again. 'That guy called back. He has asked for you to show up at the ARCO gas station at the Shenandoah Caverns exit off Interstate 81 at 7:00 tonight with USD 250. He says show up and wait. He'll get there as soon as he gets off work. I assured him there was nothing illegal done and the suitcase contains nothing illegal, but he's worried you might show up with cops. I told him there would not be a car around except yours. I told him you wouldn't ask a question and will hand over cash without opening your mouth.'

Our small bank in the town is closed. We stop by Henry and Nancy's whose business is on the way to the Shenandoah Caverns and borrow money. Racing down Interstate 81, Jennifer, Jesse, and I arrive at the ARCO station at 6:40. We wait: 7:00 passes, 7:05, 7:10, 7:15. At 7:20 an aging, once-blue pickup pulls through the station. The driver looks around and drives on. At 7:25, the truck comes back. 'Kenny' gets out—at least that is the name on the front licence plate with a heart sign above that says 'Carla'. Kenny reaches into the back of his pickup and pulls out the suitcase. I hand over the money. No word is exchanged. Inside our car, the little green case sits in Jennifer's lap.

As we leave the gas station, she opens the case and starts separating family slides from bear slides. 'In these pictures are too many memories of good times, good people, and events that will never happen again. Skull and paws can be found again, but not these deep days.'

Notes

1. With multiplying numbers of people consuming planetary resources at an ever rising rate, I offer this quaint term as descriptive of the change in the human species behaviour.
2. Johannes Krause et al. 2008. 'Mitochondrial Genomes Reveal an Explosive Radiation of Extinct and Extant Bears Near the Miocene–Pliocene boundary'. *BioMedical Central Evolutionary Biology*, 8: 220.
3. R. Nowak, *Walker's Mammals of the World*, fifth edition (Baltimore and London: Johns Hopkins University Press, 1991).
4. Gary Brown, 'Bear Behaviour and Activities', in *The Great Bear Almanac* (New York: Lyons & Burford, 1993).

5. D. Reid, M. Jiang, Q. Teng, Z. Qin, and J. Hu, 1991, 'Ecology of the Asiatic Black Bear *Ursus thibetanus*, in Sichuan, China', *Mammalia*, 55(2): 221–37.

6. Jewel Andrew Trent, *Ecology, Habitat Use and Conservation of Asiatic Black Bears in the Mountains of Sichuan China*, thesis for Master of Science (Blacksburg, VA: Virginia Tech University, 2010).

7. Christopher Servheen, Stephen Herrero, and Bernard Peyton, *Bears: Status Survey and Conservation Action Plan* (Gland, Switzerland: IUCN/SSC Bear Specialist Group).

8. John R. Napier, *Bigfoot: The Yeti and Sasquatch in Myth and Reality* (New York: E.P. Dutton Books, 1972), p. 61.

9. See en.wikipedia.org/wiki/Patterson-Gimlin_film.

10. William Bright, *Native American Place Names of the United States* (Norman, OK: University of Oklahoma Press, 2004), p. 422.

11. Greg Long, *The Making of Bigfoot: The Inside Story* (Amherst, NY: Prometheus Books, 2004).

12. Long, *The Making of Bigfoot*, p. 336.

13. Long, *The Making of Bigfoot*, p. 349.

14. Long, *The Making of Bigfoot*, p. 361.

15. Long, *The Making of Bigfoot*, p. 363.

16. Long, *The Making of Bigfoot*, p. 443.

17. Long, *The Making of Bigfoot*, p. 447.

18. Korff, Kal K.; Kocis, Michaela (July–August 2004). 'Exposing Roger Patterson's 1967 Bigfoot Film Hoax'. *Skeptical Inquirer*. Committee for Skeptical Inquiry, 28(4): 35–40.

ten

From Whence Knowledge

TWO

From Whence

Knowledge

Previous page:
10.1   A Cryptic Jungle Trail?

Source: Author

Verity of Darkness

November 1983. We need more bear skulls. They need to be sympatric, from the same habitat—at least one ground bear and one tree bear. As I land at Tumlingtar, the airline agent informs me about available return flights to Kathmandu; with the Dashain holiday in two weeks, only one seat is left, and that flight leaves in six days. If I do not take that seat, no seat would be available for a month. So I start walking. Our previous trek to Shyakshila took five days, one way.

Villagers loiter at the tea shop beyond the edge of the airfield. Over glasses of sweet milky tea at the end of the day, government officials and friends are catching up while watching the perpetual carrom game where four players flick their fingers to ricochet black and white discs into the four corner pockets on the board. From thorn bushes behind the tea shop a dove coos. I tighten my pack straps; the goal for tonight's unplanned evening hike is 10 miles away. I must cover half the distance planned for tomorrow tonight, so sleep will be in Bhotebas, 6 miles beyond Khandbari.

Then I must reach Shyakshila in two days to return for my flight to Kathmandu. For, in addition to the urgency of my airplane ticket, if adoption paperwork now started goes as hoped, a daughter will be waiting in Kathmandu.

A visibly angry young man steps out of a dry-goods store and tromps off uphill. There's no village before Khandbari, so he's probably going that far. Maybe the miles to Khandbari Bazaar can be shared. I hustle to catch up; he's walking fast and will make a good trail partner. His wife is probably beginning to cook lentils and rice for his dinner. As he turns the switchback above, I glimpse his profile which suggests he's a Brahmin. His dress suggests he is probably a schoolteacher.

I close in on him until 10 yards behind and then adopt the flow of his climb. Neither of us says anything. Westerners learn their climbing on staircases. As children we hold the banister, taking one riser at a time, each tread a stage to the next. By contrast, Nepalis learn climbing on slopes lacking symmetrical breaks, legs always adjusting with the hill. The foot placement does not get regimented with the consistency of a carpenter's cut; step selection is what makes the climb easier. Legs prefer short strides and quick movement. Ascending, bodies bob as shoulders lean in and out to the slope to balance their weight over the feet, stride adjusting, flowing uphill like water running down—that is the Nepali way, though the city-bred, staircase-taught Nepalis can also be spotted by their clothing. In a land of mountains, the way of walking speaks volumes about where the person has come from as does the accent in one's speech.

The eager Brahmin knows I am behind. He can probably describe me even though he's never looked straight at me; he did turn his eyeballs in my direction at bends in the trail. But together we've both settled into his pace. With the cadence of a shared stride, thoughts start connecting. I project his questions. Is the foreigner going to Khandbari? Probably. Where else could he be going? Who is he? Maybe Peace Corps. No, Peace Corps people carry dark green packs; his is brown. Though carrying a pack, this foreigner can't be going for a trek; the area past Khandbari is closed to foreigners.

I smile. He knows he walks fast. Twice he briefly breaks his pace, about to let me catch up. The curiosity flowing from him is not surprising: I don't fit. Nepali life is ordered like the caste you are born into. And here in Nepal not only am I out of position as a white among browns, but there is something else too, a puzzle for those who are not aware of my familiarity with these mountains. Is it the way I walk or the feelings that flow uphill from me to him?

He explores his curiosity by starting the old porter's game. I accept the challenge, and speed up my stride keeping the exact distance behind. The point of the game is to push each other to determine who cannot hold the pace. As in chess, where the king is never captured, only pinned in checkmate, the porter's game is won by wearing the other person down—and that is first shown by forcing your opponent to suck for air. When you hear him opening his lungs, you've discovered his limits. Play with him; keep changing the rate of metabolism. Sucking for air shows that your opponent is using more energy than his lungs are easily getting in oxygen. As in chess when you are up by a rook, then start snipping off pawns, victory grows by trimming away. So, when sucking starts, speed up, slow down, get the person to pant. It's like taking down a knight, then a bishop, after picking off pawns, the sign of the end game. Havoc has been caused to the body's rhythm. The victim's legs are tired, blood sugar down; the loser will soon want a rest.

For young Nepali males this is a sport the way dragging their cars on Main Street is for young, rural Western males. When you can play by the delicate rules, you are 'in'. I could concede now with some acknowledgment of his strength. Then we could walk and talk. No, this is a classic sport, let's go toe-to-toe. I open my lungs, relax my throat, shorten my stride—and make my feet fly. That's one secret: shorter strides use fewer calories than longer ones. The guy ahead is good, but he's a school-teacher, not a porter. And though he walks this hill every day, what he does not know is that I'm fresh from just having carried a 35-pound Jesse plus 20 pounds of emergency and camera gear through the Helambu Valley on a month-long medical expedition. I'm ready for a toe-to-toe, one that will also get me up the 2,000-foot climb faster. The Brahmin probably thinks the bulky pack I carry weighs 60 pounds. Actually, it is not even half of that: a loose sleeping bag, a sleeping pad, a change of clothes, and cameras.

As we cross a jumble of rocks he kicks in with a spurt that almost makes me lengthen my stride. If I continue to do so, my legs will soon use my blood's oxygen. I loosen my throat. Let the air get deeper into the lungs so that more oxygen can be pulled into the blood. The key is to hold steady my body's burn rate.

After picking up pace on a flatter grassy stretch, he slows down. I can't pass him—that's racing. My role is to always respond on the end of his string. I must now slow down too, knowing he'll accelerate again. He's gambling, thinking that I'm so tired I won't be able to recover when he speeds up. I lengthen my stride to keep my muscles working at the same burn rate, strides that have disadvantageous leverage as legs reach rather than push. I'll be ready for the spurt that will soon come.

Twenty minutes later we reach a fig tree above a level spot in the trail. A thatched tea stall stands with a rock bench for travellers to set their loads on. Turning suddenly, the Brahmin stoops under the eaves, greeting the shopkeeper who squats beside the fire and a tea kettle. He studies me as I pass. The tea stall has given him an excuse to get a close look at me. He doesn't know what he's looking for, but he knows he'll learn something as he sees me pass. Rounding the next bend, I slow my pace (glad to have an excuse to slow down); he had me on the verge of breaking.

Fifteen minutes later I hear him, each step reaching for extra distance. By the forcefulness of his walk, I guess his intent is to pass me. 'Let us be friends,' I think. So I greet him as he starts to fly by, 'Kahan jani ho?' (Where are you going?).

'Mati' (uphill) is his terse reply.

'Rati pahele?' (Before dark?) I keep my questions short to make him do the talking. Few folks walk, talk, and breathe well at the same time.

'Yes, but you won't get there before dark,' he asserts.

The fellow talks the way he walks—with a swagger. He confirms he's a schoolteacher. Every morning he walks down to his mud-and-thatch school; the trip takes an hour. Every evening he climbs the 2,000 vertical feet home; the walk uphill takes an-hour-and-a-half. This evening he stayed and played cards—and lost thirty-seven rupees. He's angry at the government worker who won his rupees. The more he talks about it, the more angry he turns at the government that employs that worker, shifting the blame to the king.

The king of Nepal is also a god, a reincarnation of Vishnu. While European monarchs established their legitimacy through divine right, in Nepal rulers became divine, both potentate and pope; a divine monarch is not just absolute but also beyond challenge. Thus, Nepal's king is the starting point of both legal and moral authority. Angry words like the Brahmin's that just blamed the king are expressions by people who now feel victimized. As these once-isolated valleys get connected with an aspiring world, such words are new, and this Brahmin in a once-isolated village is caught, like all of us, in the modern onslaught of change.

'That government worker took my money! Why must all my work contained in that money disappear?'

'Because you played the wrong card, the knave, but forgot that the king was stronger.' (The pun works in Nepali as it does in English.)

The schoolteacher laughs. 'Yes, the king is stronger. Not only does he take away my money, he is taking away my country too. Taking it and giving it to those beyond the law, even giving my Nepal to rich Indians.'

'He is not taking away your country,' I reply. 'It is his country. You gave him the country when you crowned him the king. Gifting the country to the king is part of your custom; it is a gift the people give each time they crown the king. If you, the Nepali people, are honourable, how can you now take the gift back?'

'I do not take the gift back. He gives it away. Actually not he—those close to him sell it. Maybe the king is good. These other people enter business with others and give away Nepal—our trees to India, our temple idols to Europe, our cloth to America. When we gave Nepal to the king, it was as a resource so that he could help us in return.'

'No, you did not give him the country so that he could help you. That is a new idea, not part of your historical contract. Nepalis give Nepal each time to the king because the country belongs to him. You are his people; you belong to him. What he wishes must be your wish too. What he wants, you must provide. He is your king and your god.'

'Who are you?' the schoolteacher asks. 'Why do you talk like this? I only speak how your man Thomas Jefferson taught me to speak.'

I reply, 'Why do you say "my man Thomas Jefferson"? Jefferson is an American. I'm a Russian. I'm ...'

'Then your man Lenin,' he replies, unfazed. 'I don't care who you are. I talk now about people taking what is not theirs. You, Mr Russian, like all who come, take from Nepal. And what you give back are playthings for you. Just yesterday a French tourist, as she waited for her flight, walked to my school and tried to talk one of my girl pupils into going to France with her. She gave this girl a bag of bonbons and asked her name; then while the girl's mouth was full of candy, she asked her parents if she could take the girl to France! Does she want a daughter? Or she wants a slave? Does she think her life is better than ours because it allows her to have bonbons? You all seek to take. You Russians are trying to ...'

'That is why you need the king,' I interrupt. 'Who else can hold Nepal together? Who but the king can bring a Brahmin like you here in eastern Nepal—a man with no fields because you are a younger son cursed with big ideas, a man who so desires money that he gambles even when it is illegal—together with a Brahmin from western Nepal, especially when you two do not even speak the same language? Who else but the king? Even your gods can't, for despite being Brahmins you both don't practise the same rituals.

If two Brahmins are unable to get along, how can a Sherpa from the mountains get along with a Tharu from the flat jungles? Those Nepalis eat different foods, cannot talk to each other ... but still when the sun comes through the dark night the next day, both respect the king. If Nepal ever enters true trouble, only the king will be able to rise above your hills and differences with the authority to act for national interests.'

He is stunned. 'How do you know so much about me—that I'm a younger son, so much about Nepal? You must be a spy. How do you know I'm a Brahmin? How do you know I want money? You're a spy!'

'I'm no spy. I was also raised in these mountains. My Himalaya mountains are in India to the west of the Nepal border, and I often come back to these mountains as a friend. I know you just as you now know the ideas of the West. Knowledge flows both ways in today's world. Therefore, in this world with ideas that flow, we are brothers. I know you are a Brahmin the way any Nepali would. The shape of your face tells me, as also the string hidden under your shirt. You have no fields or you would not be a schoolteacher, but you have a good education, so you must be a younger son of someone who had fields. Of course you are hungry for money. Why else would you play cards?'

He laughs. 'You are a brother of Nepal, dear Russian.'

I wince. With this fabricated layer to our conversation a new darkness has grown around us, the world of deception. The schoolteacher enjoys such a debate. In villages like his, among those with book learning, it is popular now to try on ideas they've read, even talk about revolution. In these isolated villages debate among new ideas and proposing new structures, like the changing seasons, gives hope. Now, though, started with a glib wordplay my Russian ruse could backfire on me.

Politics in Nepal is more complicated than the picture the Brahmin paints. Players constantly change, not only in terms of their levels of influence but also in their colours. There is no black, no white. Shades of grey merge with the shifting influence to make politics impenetrable like the night covering the trail before us in darkness. The game of moving pieces goes on, but it is more than multidimensional chess. Politics heaves up and down like the Himalaya themselves. Here there's no level-playing field, for the legal system in this fractured land functions without precedent, run totally on monarchical wish, and the king, as I know well, has a mercurial temperament.

Our pace has slowed. Neither of us can see ahead. As we talk further, ideas lead to bafflement. Clarity as to where to go in social direction probes as my feet do about the trail. In both we learn from footfalls of those ahead, and from the lighter line of packed earth walked. In going through the dark, whether in life or daily trails, we find the path by walking it, discovering by feeling and not by seeing, ready to lift feet when there is no firmness beneath them. Answers come sometimes when no questions are asked, and questions linger pretending to be answers.

Our strenuous walk has relaxed barriers of the mind, opening the mind to the night. Like the liberating high of long-distance running, when the body falls away and senses float, awareness opens. 'I don't like your Soviet bear,' the schoolteacher states abruptly.

'Why?' I reply absently, concentrating on the trail, now on the back of a ridge where the moonlight has not reached. Around the next turn bamboo rises above the trail against a sparkling black sky swooping in from up high.

'We Nepalis never trust the Soviets. We hear you talk about how you help people, but we know you do not. Maybe, as they say on the Voice of America, you seek to control the world as you control your own country. My cousin brother thinks you act big because deep within, you know you are small. You possess a big country, but it is filled with emptiness—you know that, and that is why you keep pretending.'

'You are wrong,' I reply. 'My country helps people. We help by training Nepalis in Moscow. This year we sent forty students to become doctors, twenty students to ...'

'Don't tell me about your Soviet doctor training. Another cousin brother studies in Moscow. Last month he returned for his father's funeral. He told me that your training never permits him to touch Russian patients. Hah! You show him surgery only on TV. Never is he allowed to conduct a real surgery, except on animals. Never does he meet Soviet students, especially girls. Is that useful education?!'

'But we are training doctors—that is helping Nepal.'

'How can you say you are helping? You and the Americans play games with us. I have another cousin brother who works at the customs office on the Indian border. Two months ago trucks arrived loaded with boxes addressed to the Soviet ambassador. The inspector demanded that the boxes be opened; this cousin brother heard the argument.'

'A customs inspector can't open boxes addressed to an ambassador! That is a violation of diplomatic privilege.'

'Our customs have the privilege of Gurkha soldiers! When Gurkhas suspect trouble, they do what's necessary. The boxes were opened. They were full of electronic instruments, big antennas, expensive equipment. I have another cousin brother. He is a secretary at the American embassy. He, too, was at the funeral where we were talking. He said the Americans are laughing. The Americans say that the Russian equipment is for listening to telephone conversations, maybe even within the

Royal Palace! You do not allow Nepalis to private talk even in their bedrooms!'

We've arrived at the police check post; Khandbari Bazaar rises before us. A sentry calls, 'Halt!' This Russian is going to look mighty suspicious trying to walk past a check post at night with his flashlight off. To walk further, special permits are required. Knowing where the post was, my intent had been to walk around, but deep in the conversation I'd missed seeing the post. This Russian is going to look very much like an American spy when he shows his American passport in front of the Brahmin teacher.

'You walk on while I fill out the papers,' I say to him, worried that I might not even make it out of jail and to Tumlingtar to catch the plane—let alone to Shyakshila.

The Brahmin calls to the sentry. 'It's OK. I'm Ram, the school-teacher, walking with a traveller. We're talking.'

'*Bistari janos*,' (go slowly) comes back from the sentry as in the dark it is not evident that a foreigner is also walking alongside.

10.2   An Evening View of the Arun Valley

Source: Author

We enter the bazaar. A small restaurant is serving *dal bhat* (rice with lentils). Ram says goodbye. I place my pack against the earthen wall. In the middle of the room a hand-hewn wooden bench sits in front of a store-bought table. Others wait; some at the table, some sitting by the wall, talking and drinking tea, not yet eating. Only when the conversation is finished do Nepalis eat. The strict Brahmins among them will not eat, but will leave after sharing tea with the others. But with 6 miles to cover to reach Bhotebas, I hurry through my meal. The goal remains: reach Shyakshila by the day after tomorrow.

Can the Priest Be Trusted?

Two days later, as the evening shadows climb the valley, I walk the cobbled high trail leading into Shyakshila. The doorways are locked. Only chickens are on the walkways, cackling as they scatter in front of me, pebbles emptily clattering under my feet. Even the pigs are not rootling.

Then a hum—to the right and down the slope. A child's head pops up from behind a stone wall; the child runs away towards the sound. As I follow, a shout rises off the rock walls, and a crowd of hundreds comes towards me. People roll like a wave up the shore. Behind the crowd, in an open square are hundreds more. As the crowd closes in, phrases pass: 'It's father of Jesse!', 'Where is Jesse?' In the middle stands Myang, smiling ear to ear. Then breaking from the crowd he approaches me with head bowed and palms pressed together. He grabs my hands and touches them to his head. 'Where is Jesse, Jesse's father?'

'In Kathmandu with his mother,' I reply, looking at the sea of people. 'I've come alone.'

Myang pulls me through the crowd. A temple rises with walls of freshly hewn grey stone. Atop glistens a tin roof; the only such in Shyakshila where other roofs are of woven bamboo. A year ago there were houses in the centre of the village. Now there is a temple! Light reflects from its new tin.

'Where did this come from?'

'This is our new temple, Jesse's father. We built it. The king's government gave us 80,000 rupees. More than a school, water system, or health post, what our village needed to engage the spirits of the valley was a temple. So with the government money we built this!'

Approaching me through the crowd is the lama, a man who often came to sit by our fire. Two boys flank him as his assistants.

'Welcome to our celebration, Jesse's father. We were waiting for you,' he says formally. 'We heard yesterday that the guest we were waiting for is a white person. Come sit inside the temple.'

'What? Waiting for me?'

'Yes, we first decided to have this celebration after the cement to build the drain gutters arrives. Until now, we used no cement. When the cement arrives the temple will be finished. But my brother, a lama who can study astrological books, had told us two weeks ago that we must celebrate today. He said a guest was to come. We did not know who the guest would be. This morning when we gathered, news came that the guest was a white man. The people here have been waiting for you today. Thank you, Jesse's father, for coming.'

News of my arrival reached ahead of me? I thought I was the fastest walker on the trail. Open ditches wait for their cement. The lama waits for me to follow him inside, but I move to a pile of left-over rocks on the terrace. It will be smoky inside from yak-butter lamps and people packed into what I guess must be a 16-by-16-foot room. So, adjusting two rock slabs I make a seat, then two more to prop my pack to allow a mesh-woven backrest. If I'm to be the chief guest, I should settle in.

Was the date switched two weeks ago? Two weeks ago when the lama's brother said that, I hadn't bought my air ticket to Tumlingtar. In fact, it was two weeks ago on the medical expedition that I made the schedule adjustment to come here. How could people know this about someone in another Himalayan valley and predict the date of his arrival? They so trusted this knowledge that hundreds of people are here today waiting for me.

Did I misunderstand anything? It would make sense if I just happened to show up for their celebration and they were happy. But they changed the date. I pull out my water bottle, and as I casually drink I ask from others the story. In a few moments perhaps I could request for boiled potatoes to get my gut ready for the home brew that the chief guest will be pressed on to drink.

The people have opened a space in front of me, shaping it like a stage. On the other side of the packed earth is the temple. Precision holds together the 2-foot thick walls; no mortar has been used and each rock is snugly packed like the people crowded around me. Below the main temple roof, about halfway down, flares another roof, making a porch on all four sides—a roof that not only gives grace but will also flute out the monsoon away from undermining the building. Which architect designed this? Certainly no villager ever designed a temple before. No nearby temples share this design. The shape came from lines etched in someone's mind; its construction came about because others trusted the one who was etching the lines.

But there is no doubt among these people. I am frightened, for soon they will lose the trust acquired over millennia. They built a temple because they trusted 'a way to engage the spirits of the valley'. Soon that may disappear as people forget the inner wealth they now have, that their ancestors have grown. The confidence, pieced from lived experiences, and the knowledge held across landslides, epidemics, and wars will come unglued. They will start believing that the outsiders will come, or the king will give them something. It will not be long before they learn to trust the telephone, believing in a call from folks like me saying that I'll be there in two weeks. Had I so called, not knowing the flight schedule, expecting to walk more slowly, I would have foretold that I would arrive in two days from now. Did that astrologer also predict that all the aeroplane seats are booked?

ᏒᎳ

WOMEN GATHER IN A ROW ON THE STAGE awkwardly and shyly. With the cadence of the accompanying song, they start shuffling meekly. One by one, men drift in behind them, joining their rhythmic movement and making clear, suggestive remarks. What started as a song swells to a chant. I am still wondering how they knew that I was coming—discomfort rises as I fathom how men *and women* are treating the women on the stage, both men and women giving gestures of the women. I cannot understand their Loomi language, a variant of Tibetan and Nepali. Only when talking with me do they speak Nepali. However, you need not know a language to understand boldness. In these isolated mountains, people are more sexually open than in southern or central Nepal. Behind me a young girl turns with a short shriek and hugs her mother's legs.

In the dancing queue a girl wearing a pink blouse blushes; she might be fifteen, maybe sixteen. Her once-olive earlobes now turn crimson; titling her head she rubs a hot ear on one shoulder and then on the other. Lines continue to shuffle. The crowd chants faster and the dancers pick up pace. A man slides down the men's row and steps directly behind another girl in a green blouse. He calls out something. The crowd snickers. The lass in the green blouse looks at the ground and blushes. Is the girl in green a younger sister of the one in pink? The man calls again; the crowd snickers more. Men and women seem to view this as fun—but not the girl in green.

The man calls out again; the girl in green bolts, rushing into the crowd. He charges after her and from the crowd comes a cheer. The crowd arrests the girl's flight, and she is spun around and thrown back on to the stage. The man catches her. She hangs handcuffed by his grip, her head bowed and shoulders limp.

'Take her! Take her!'

Straightening in his arms, she stiffens before the man, shoulders back, more erect. There is fire is in his eyes and mettle in her stance. She steps back, but her foot hits a rock. She slips from his reach, struggles to recover, and falls to the ground—arms out, legs apart, on her back spread-eagled. He lunges forward.

Bodies surge around me. Doesn't this girl have a brother or father to protect her? Anywhere else in Nepal a man would come forward. But maybe not. There is something everywhere about groups of people committing a wrong—regardless of their education or position or culture—that incapacitates even the well-intentioned. It takes uncommon strength to buck shared sin. The victim-to-be, the girl in green, lies knowing what will come in seconds. After it's over, we call them martyrs, but who steps forward while it is happening? The temperature around me has increased. The sounds seem like laughter, but I wonder, in reality how many of them are crying.

'Do it! Do it!' someone screams.

Coming down the man lands on her, presumably to snatch her away from the crowd. But he was moving too fast. Was it an enthusiastic push? With his fall, he slides forward on top of her with his stomach over her face.

She bites hard. He gets up with a howl, grabbing his stomach through his thin shirt. He's off, legs flailing, running into the crowd and holding his gut.

Fifteen minutes later, the villagers still hold each other, weak from the hysteria as they retell the story. That man will be asked about his stomach for the rest of his life. His name might even change to 'Stomach'. In time, the dancing resumes, looking as if it will continue through the night.

I leave the party at eleven. Chanting and prayers continue in the temple: 'Om mani padme hum, om mani padme hum ...' (O jewel in the heart of the lotus). On the hewn-plank floor of Myang's house nearby, I unroll my sleeping bag; it was a long day. When

I wake up, the dancing and chanting still continues outside. As I open my bag the sound of the zipper awakens Myang's wife who is lying in her bedroll on the other side of the room. She kindles the fire from its embers as I watch. As she crosses the room minutes later to bring me a glass of tea, I notice that Myang is not around; he's been with the party all night.

I go out onto the porch. The sun has not risen yet over the ridge; the morning is light grey and cold. Sipping a second glass of tea, I watch women head to the edge of the village where they can privately perform their routines, watch children move to the fields to start their chores, and see men stagger home. As the village comes to life, I let out the news that I'm looking for ground-bear skulls.

A long talk follows with Lendoop. 'It is dangerous to kill ground bears,' he says, 'and there are no skulls in the village right now.' And he tells me how aggressive these bears are and how big they can be.

As the day goes on, I am told that even tree-bear skulls are not available. Everyone's denying that they kill the bear, indeed both bears. But when the night falls, out of the darkness people come and speak of how months ago a new government rule was announced, according to which they can no longer kill bears. And in the protection of the dark is exposed for my inspection the guardians from grain storehouses. They explain how it is okay to sell because they will kill new bears and the vermin will get to know the new skulls. To worried but eager sellers I explain that I have permits. The villagers don't care about the international treaty on trade in endangered species; what worries them is that the police post is a half-a-day's walk away. A few months ago in Hatiya village, someone was arrested for selling bear gall bladders. When I leave Shyakshila at predawn the next day, in my pack are two more tree-bear skulls that are clearly not juveniles. My permit will allow only two skulls. I'll return next year for the ground bear, and to get its skull they say that I'll have to go to other villages.

There are sixteen hours each day from predawn to full night. I work backwards from my flight ahead in Tumlingtar, setting times to places, planning a pace that I shall force myself to keep. Two passes must be crossed, one with a 4,000-foot gain. Four more hours are available in the two days ahead as compared to the time it took me while coming in, so I plan an hour of relaxation at the top of each coming pass.

I wrestle with puzzles that have no real answers. What is the relationship between an American man and a Nepali village? What are the pressures bears in the Barun Valley face from Nature and from men? The Convention on International Trade in Endangered Species, did it get to these valleys because of law enforcement or someone seeking to break the law? In secret markets in these hills, what are the other unknowns beyond bear gall? I know a single musk-deer pod sells for USD 50 in Khandbari and also that it can be further sold in Hong Kong for USD 500. A man exploiting a girl, a bear exploiting a man's cornfield—what are the driving forces in these challenges as well as those across the earth?

For me, an American, it is tempting to consider these to be issues extending across the oceans—but growing in the shade of hardwood trees in my USA home is the ginseng root, which is in demand as a sex elixir in the same Far East markets that sell gall and musk. My mountain neighbours sequester ginseng patches for these markets where the plant is sold legally. But they also illegally sell the gallbladders of bears chased down by their dogs. The taxidermist back home was taken in four months ago and his shop shut down. What is the difference between my town and a Nepali small town in global systems?

However, how did 1,000 people know that I was coming? How did a fact that was unknown to me even until three days before become known to them? Some patterns take shape beyond the realm of our knowing, yet are known to others. How did that squirrel now on the tree limb above know, as it darted off, that a hawk was overhead which I could see while the squirrel appeared to be looking at me. Or, was it happenstance that the squirrel darted? Tree bear, ground bear, Yeti—what are the communications among us of our knowledge about these?

Two passes and fifty hours later, I arrive in Tumlingtar three hours ahead of my flight, hours to stretch having tea in the thatched hut watching the carrom game. On entering I placed a special order with the shopkeeper for his fried bread, and he could have made it then. But the time was not right; waiting to make the dough, kneading it, and letting it rise. So now, three hours later, having waited until his customary time came, he fries up the bread. As it sizzles, the Twin Otter drops down from the clouds and buzzes the field to disperse the goats. With both my hands filled with greasy bread I pass through the bamboo booth that is the security check. I learnt eight months before that this bread, when hot and sprinkled with bazaar sugar, is tastier than a Dunkin' doughnut.

As the aluminium capsule ascends into the clouds, Myang's words return to me, 'More than a school, water system, or health post, what our village needed to engage the spirits of the valley was a temple.' Myang lives in a world that believes in the power of temples and that these must be prioritized over schools, water systems, and health posts. The theologian Karen Armstrong helpfully clarifies this:

> We tend to assume that the people of the past were (more or less) like us, but in fact their spiritual lives were rather different ... evolving two ways of thinking, speaking, and acquiring knowledge, which scholars have called *mythos* and *logos*. Both were essential; they were regarded as complementary ways of arriving at truth, and each had its special area of competence. Myth was regarded as primary; it was concerned with what was thought to be timeless and constant in our existence.... Myth was not concerned with practical matters, but with meaning ... [which] provided people with a context that made sense of their day-to-day lives.[1]

Myths are not to be taken literally. Literal interpretation removes their ability to bring the obscure into understanding, to show their dimensionality. The obscure is not only inside a question but can also be outside people, giving perspective of collected of events not quite clear in the complex world. In contrast to myths, logos pushes understanding into a logical order. Again, Armstrong says:

> *Logos* was the rational, pragmatic, and scientific thought that enabled men and women to function well in the world.... Unlike myth, *logos* must relate exactly to facts and correspond to external

10.3　Bear Skull and Paws for Sale in Shyakshila

Source: Robert L. Fleming

realities if it is to be effective.… Unlike myth, which looks back to the beginning and to the foundation, *logos* forges ahead and tries to find something new: to elaborate on old insights, achieve a greater control over our environment, discover something novel.… *Logos* could not answer questions about the ultimate value of human life. A scientist could make things work more efficiently and discover wonderful new facts about the physical universe, but he could not explain the meaning of life.[2]

Riding the plane back to Kathmandu through the cumulus of ideas, I understand how these ideas of myth and logic are at play in the skulls I carry. I went to Shyakshila seeking a logical explanation regarding the two bears; on my arrival I discovered a mythical explanation for my trip. Of course, neither explains the Yeti—but the Yeti grows more strongly as a being which speaks powerfully as a myth and also a scientific fact. Mythos in Nature is wildness. Logos in Nature is science. In each of these are worthy, truthful meanings.

Truth from Intuition

Back in Kathmandu, I walk up behind Jennifer who is talking to Jesse in a hotel garden, 'Today, Papa leaves Shyakshila. Remember the village where all the people looked at you?' She turns startled; no astrologer has told her of my arrival. She skips asking me why I've returned so soon and starts her story. Two days earlier news had come that paperwork had cleared and we could adopt a Nepali orphan.

'Jesse and I went to see this girl. The orphanage matron does a splendid job, but her resources are so limited. A large closet serves as the nursery. That's it—a large closet. At one end, a window lets in light. There are baskets lined up in two rows on the shelves against the wall. Each contains a baby. Picture a walk-in closet and baskets of babies rather than clothes on the shelves. Not a single baby was making any noise.'

'A space so small will probably be warmer in winter, Jennifer,' I insert.

'As we walked into that room, Jesse held me tightly—he, just like me, sensed something. Mrs Shrestha, the matron, pointed out our baby. She's lovely, three months old. But at my first sight I felt that this baby's not mine. She is like a picture far away, not part of our family. Lifting her from that basket was like reaching across a chasm. I'm sure she will grow to be part of our family, but now we're going to reach the other side of the world.'

Just having walked from Shyakshila, my head spins. We pushed hard, Jennifer and I, to get our papers prepared in the US to have our second child an adopted one. Is Jennifer's sixth sense making her reject that baby? Is this her way of telling herself (or me) that she doesn't want to adopt?

'We'll not take the girl,' I say. 'I'm going over to tell them that our request is cancelled.'

'You can't do that, Dan'l. They've made a lot of effort to arrange everything, and done so fast. An orphan who is alone in the world has been assigned to us. We're her parents now. You're her father!'

'First, you say she isn't ours and then you say we must take her. You've been thinking about this for two days—and you're still not sure? That says a lot. What am I supposed to do?'

'It's worse than what I've told you. There's another girl in another basket. Walking into the nursery my heart went to that girl. This one, two baskets away, felt right even as I picked up my own. As I entered, this other little head turned straight at me, a thin smile on her sweet lips.'

'What?'

'Yes, I asked about her. Mrs Shrestha said she was taken; she has another sponsor. Oh, what a sweet smile she had.'

'Maybe we can have the girl changed. If you like one baby more than another, I'll talk to them.'

'Stop being like that, Dan'l. These are babies! It's not like exchanging clothes. Didn't you hear that the other girl belongs to someone else. You don't just take someone else's baby.'

'OK. OK. But if you don't feel right about this, we're not going through with it. I'm going over to talk with them.'

Reaching the orphanage, I follow Mrs Shrestha down the hall. I mention how many other people want a child. Turning into another corridor in the old refurbished palace, I say that Jennifer and I are unsure about adoption; maybe we still want another of our own. Mrs Shrestha leads on. The nursery is indeed an old closet that once held the clothes of wannabe royals. Baskets line the shelf, looking like parcels of clothes; no noise from any. I step over to one. This girl's head had turned as I entered. I run my finger across her cheek. A smile flicks up, lips tight across her face, her eyes intent. There's discipline in her body. Then I notice how thin that body is. Undernourished, maybe even malnourished? She has no hair. Her smile flicks again. Are those dimples? I pick her up while Mrs Shrestha talks with the matron.

'Your girl is here,' Mrs Shrestha interrupts, pointing two baskets beyond.

My face flushes as I put the first baby back. The girl I turn to is better nourished, the skin of her face looser. She gives a strong kick, smiles widely, kicks again—is that kick her trick? But as I cradle her, I feel that Jennifer is right: this baby does not feel like ours, as right as the other one, yet I know love and time will bring her close. She kicks again.

'What do you know about our girl?' I ask of Mrs Shrestha, 'How was she orphaned?'

From the other room Mrs Shrestha brings the file: 'Girl. Born September fifth, caste Maharjan. Mother, aged thirty-two, died from haemorrhage during childbirth at Thapathali Maternity Hospital. Father, who has no relative or other children, turned the baby over to the hospital saying, "If I care for the baby, my fields will die. If I care for my fields, the baby will die." The girl arrived here six days ago, very underweight. She needs special care; that's why we think you are the right parents, and because of her malnourished condition you must take her immediately.'

Mrs Shrestha looks at the plump baby I hold, then again at her file. She looks again at the baby, then back at her papers. She calls for the matron. The two go to the matron's office. Another woman joins them. The three talk excitedly in the next room. They come into the nursery, look at my little girl, then go through the other baskets. They're talking fast in the Newari language, and I can't follow. In my arms, our baby starts crying. I rock her and walk around the room as she continues to cry. Mrs Shrestha comes back holding two files and goes to the basket with the girl who has tight skin and lips. Looking in, Mrs Shrestha is visibly distressed. The other two women stand close, talking softly.

'It must be. It has to be,' the matron says nodding.

'Yes,' says Mrs Shrestha. 'But how did it happen?' The matron takes my baby, and our little girl quietens immediately. Throughout, other babies in the room have been quiet.

Pointing to the first basket, Mrs Shrestha says, 'We made, I think, a bad mistake. That is not your child; yours is malnourished. Your child is this one.'

I look towards the first basket. Is she the same girl Jennifer was attracted to? That tiny baby with tight lips looks too weak to smile, but she does. Yes, those are dimples. I look at the women and understand. It's like Gilbert and Sullivan's *H.M.S. Pinafore*: they mixed up the babies!

'Wait. I must get Jennifer,' I say and rush out excitedly.

Back on my motorcycle I race through Kathmandu's streets, the thumping pistons of that old BMW whining as I twist up the throttle and coax seconds off the trip, snaking the big bike through the crowds, accelerating on the open road. Collecting Jennifer and Jesse, we weave back to the orphanage with Jesse astride the gas tank, holding the handlebars.

Mrs Shrestha explains, 'The well-nourished girl belongs to a Nepali businessman. A-month-and-a-half ago, he was on a bus with his wife and three children on his way to Kathmandu to see his relatives. On a steep turn the brakes failed and the bus went over the cliff. His wife and children died. He was one of the few who survived. Lonely, he walked through villages looking for someone to call his family. He found this girl who, for some reason, also had no family. Now she is his. He left her with us while he arranges for a woman servant to care for her.'

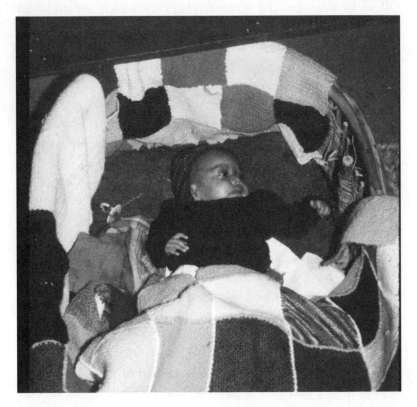

10.4 Our Daughter Tara at the Orphanage—Smiling to Come Home

Source: Author

The matron picks up the thin baby from the first basket. 'This girl is yours, the one that needs urgent care. We think the mix-up between baskets happened when the hospital people came last week to immunize all the babies. All were out of their baskets then and crying because of the needles. This little girl whose mother died from bleeding is your daughter.' The woman places the tight-lipped girl in Jennifer's arms.

୧୬

CHANCE—WHAT IS IT? HOW DOES IT WORK? There is a fable of a man who fired gunshots at a wall. After emptying his gun he went up and painted outwardly concentric circles around the cluster of bullet holes, claiming that by having the cluster inside his painted circles, he had perfectly hit the centre of the bull's eye.

Do we sometimes create our own accuracy? Was there a real telepathic exchange between the Brahmin and me while walking uphill with our minds truly sharing the same mental groove? Might my knowledge have been purely deduced from the shape of his nose and the way he was walking, assembling prior similar conversations and putting a circle around those points?

The Shyakshila village changed its inaugural date; that was certain. However, was doing so true precognition? The lama did not predict my arrival; he had just predicted the arrival of a guest. Coincidentally, I reached the village on that date, but it could have been anyone, or the astrologer having said a special event would come then presented as evidence perhaps a remarkable bird that landed on top of the temple, or something else. Such signs then would have been validation. The lama would have to make multiple such predictions to know for sure.

A child's telepathic exchange with her new parents—is such a thing possible for a baby? Astronomer Carl Sagan spoke of the ability of infants to describe events and places in prior lives; not that he was certain of its veracity but he claimed that their accuracy was so uncanny as to warrant scholarly study. Our new daughter's communication, Jennifer-to-her and me-to-her, is the most convincing of these three astounding occurrences—for the awareness came separately to both of us and so firmly that neither of us had any doubt. Science might not accept such evidence, but it is so strong that parents act (and are willing to die) on its basis.

Supra-understandings have been happening, as best as can be known, since the dawn of human life. Knowledge that is not conveyed through the five senses—but is somehow transmitted—has come too many times, filling lives of communities or of parents knowing something terrible (or wonderful) about their children. Such knowing does not counter science. Science does not claim to be all-knowing; it is only error-catching and this distinction is important. What is striking about this process is how such knowledge also dawns upon the hardboiled sceptic who insists on working from facts—then eureka, an answer arrives totally unrelated to the scientific processes the scientist had been following. Science does not explain this, but assigns it the name 'intuition'. Evidence itself did not create the knowing because data was not yet collected; also the hypothesis had yet to be postulated. Only the idea passed—from that, knowledge was reverse engineered—then proven correct by backward calculus.

With our daughter, the reason that drove our knowing could have been anything from sympathy evoked by her malnutrition, to the smile that, at some level, was a family feature, perhaps a feature that our son Jesse already had. Or, it could have also been some family bond speaking across genetics of global difference. Whole cultures are convinced of reincarnation while science has yet to prove it.

In the three instances in this chapter, I possibly picked some random coincidence. Perhaps I appear to be developing a line pointing to extrasensory communication because I am drawing circles around random bullet holes. But then, maybe something is communicating with us about which knowledge has not been discovered as yet. Perhaps thoughts radiate using waves or chan-nels yet undiscovered; these are then partially picked up by our partially developed senses, the way some people have more sensitive ears or eyes. Because a sixth (or more) sense has not been defined yet, who can guarantee that humans have only five senses?

For Grandpa and Jim Corbett, their 'sixth sense' saved their lives in the jungle on repeated occasions. And because their lives had been saved, both men believed in it. Perhaps what alerted Grandpa and Jim Corbett was not a 'sixth sense', as these two men were deeply attuned to jungle messages. But what they were gathering was a call that a bird some distance away had made and was then picked up and passed by another. Information may be coming, and perhaps the explanation is that we are improper in its attribution.

We walk on paths in life that more than half the time are dark-ened trails to the unknown. We seek light. We enter communities of living where events are happening but we do not know why. For example, our children always surprise us with new traits—from where do these actions come with some genome imprinting to that specificity? We are individuals, but beyond ourselves exists a great cosmos of both that is known and not known. To many people, such a cosmos feels as though it is there, the way we felt about our daughter.

One striking feature of science is its certitude. But today's science may be wrong simply because it is so sure. It looks for cause and effect, seems to identify connection, but the world it operates in is complex with dynamics beyond those being defined. What may be a cause today will not have the same effect tomorrow, because the interdependencies around the relationships between cause and effect relate in new ways on a new day. How can it then be a cause—rather, it was a one-time explanation. There is a conceit in science (at least in its certitude) that is as blind as the conceit in religious belief, both of which postulate truth. Questioning is always valid, and certitude must always be suspect.

When science is humble, then comes value—just as there is value in religion when it sits in awe, not judgment (when it does judge, it usually commits a grave sin). It seems that under-standing comes when religion is humble. In the words of Søren Kierkegaard: 'Life must be understood backwards. But it must be lived forwards'[3]—the corollary to this being that in going forward we walk into mysteries where explanation comes to us after the experience.

Each age will inevitably have its respective disciplines of alchemy, for each age has its conceits grown from false starting points in which people deeply believed in those times. In them, adherents become certain when actually they are incorrect. Proto-scientists in the Middle Ages developed theory, terminology, experimental process, and laboratory techniques, but their assumptions were flawed, for gold cannot be created from lead. Robert Boyle, who developed the science of chemistry, began as an alchemist. His brilliance is that he learnt—the method worked, he concluded (and because of his questioning we today have science), but the assumptions he was certain of in the beginning were wrong. Similarly, the twentieth century believed societal systems could be planned, and such planning would create productive societies. But the societal-planning discipline crumbles (to wit the Soviet Union). Similarly, the financial world believes its numbers describe global transactions, but they cannot predict an economy (to wit the missing prediction for the Great Recession of 2008).

In life are there patterns we do not detect yet? Or, are random events simply being grouped together? We know not what we know. We walk down trails in the dark, feeling for our route when not seeing it, getting messages though we do not know how they come to us and from where. But we are always sufficiently certain of the patterns of life that each day we get up and start anew—for from somewhere knowledge comes.

ⵣ

AS JENNIFER HOLDS OUR CHILD, A TEAR FALLS on to that olive-brown cheek. And then, a family of four arrays itself on the big black motorcycle. Jesse again sits astride the gas tank holding the handlebars. Side-saddled behind me, Jennifer cradles our daughter in her lap—the little Tara protected by a new mother's arms. In an-hour-and-a-half, we're due for dinner with friends. Before that, we must track down diapers, bottles, and formula from some shop in the Kathmandu bazaar. A little girl has a lot of growing up to do. Several of Jesse's T-shirts will serve as Tara's new baby gowns.

It is 1983, Thanksgiving Day.

Notes

1. Karen Armstrong, *The Battle for God: Fundamentalism in Judaism, Christianity and Islam* (London: HarperCollins, 2005), p. xiii.
2. Armstrong, *The Battle for God*, pp. xiv–xv.
3. Søren Kierkegaard, *Journals IVA 164* (1843).

eleven

The King and
His Zoo

December 1983. We need a reference skull from a ground bear for our three tree-bear skulls. According to science, there are ground bears in Nepal, but I can find no museum anywhere in the world that has a *thibetanus* skull from Nepal, whether of tree or ground bear. So I am meeting my friend Shah Dev again.

I wait in the antechamber. Each visit is always a bit of a surprise—will he be the friend with whom I once partied, or will he be aloof with the dual supremacies of being the king and the reincarnated god? He has been both, sometimes on the same visit. The door of the antechamber opens. Energetically, His Majesty's principal private secretary motions me in.

As I follow him down the hall, I ponder the meaning of being a king in the modern age. A question which, perhaps, only kings contemplate, and only a few are left. The problem is, if one is an absolute king, the position comes with no helpful checks and balances but with the unchecked hopes of people. The position does not allow a mistake. The question grows more challenging when the king is also a god—for gods are presumed to have ultimate knowing. Absolute authority coupled with sanctified guidance would be splendid, if it did not have to be carried out by humans.

Such leadership is flooded with requests—a business permit, a treaty proposed by a government, even an old friend who asks about a bear. And in all of this, kings survive when catious—and caution and protection are made easier by isolation. So they surround themselves with walls (of many kinds). But there is an undermining problem with living in isolation: it is easy to make mistakes.

The man who leads me to this cautious king is the loyal servant who opens doors, moves between rooms, and I've learnt that he does not like to be surprised. As the private secretary and I hurry down the hall, I spot the minister of finance in an antechamber reviewing his papers for his audience that is scheduled in fifteen minutes. Yet, as we approach the audience chamber, I know my role over the decades with regard to my isolated friend has been to bring him unvarnished news—so we often talk for hours. But this time, the private secretary has just told me: 'You absolutely must take no more than fifteen minutes. I've fit you with some difficulty, no old school talk.'

As we arrive before the door, behind which I know is a massive formal royal study, standing guard is His Majesty's aide-de-camp, who smiles in recognition, and with a polished movement swings the door wide open.

The king of Nepal, His Majesty Birendra Bir Bikram Shah Dev, age thirty-nine, is positioned in the centre of the room, feet astride a snow-leopard skin, hands clasped behind his back, grinning as I walk in. He pumps my hand warmly. Forty straight-back chairs line three walls, the fourth wall presenting, alone, by itself, His Majesty's desk. In front of the desk, grouped around the snow-leopard skin, are three armchairs and a coffee table. The room's formality and its host's regal/divine position disappears as two friends settle into the armchairs. Shah Dev continues to grin, sizing me up. I do the same, noticing that he's lost weight and trimmed his moustache. My mind goes back to the schoolteacher in Khandbari who speaks about his king without knowing him.

Shah Dev reaches for his tiny pipe in its stitched leather casing and picks up the blue tin of Dutch tobacco, custom-blended and shipped to the Himalaya from the shores of the Zuyder Zee. He scrapes the pipe bowl with a little tool made of stainless steel that is used even by commoners, flips the tool around, and packs his tobacco. As the tool flips, I spot the Harvard crest in the steel and the word 'veritas'; it is the same tool he bought one afternoon when we stepped out of the rain into Leavitt and Pierce just off Harvard Square. His pipe lit, Shah Dev looks at me intently.

'So, Dan'l,' he says with an emphatic nod, 'what have you been doing?'

'We followed your suggestion, Shah Dev, and searched the Barun Valley some months ago. You had said that no place is as wild in the whole of Nepal.'

'What did you find?'

'Well, villagers report two bears in the jungle. Both are black. One is large, very aggressive, lives on the ground, and when dead, even five villagers will have difficulty carrying it. This bear seems to be what science calls the Asiatic black bear, *Ursus thibetanus*. I'm sure you know it; you must even have shot one. But villagers also report another bear, which, for us, is new. It is smaller, shy, and lives in trees. Two villagers can carry a dead one. Villagers call the big, ground-based bear bhui balu and the little, tree-living bear is called rukh balu. We found evidence to support the villagers' claims—a bear does live in trees and makes unusual nests. We found five such nests, a trail of footprints in the snow, and possess three skulls. As far as we can determine, the rukh balu has never been scientifically described.'

'What do you mean? The rukh balu is not unknown. It is found in several places in Nepal.' Shah Dev looks at me quizzically.

'You mean Nepalis generally speak of the rukh balu and the bhui balu?'

'Not every Nepali, or even half of the Nepalis, but those who know about bears. Or maybe I should say the Nepalis living near the jungle know of the rukh balu.'

'Shah Dev, this Nepali knowledge is not scientific knowledge. Science says there are four bears in the Himalaya: sloth bear, *Melursus ursinus*; red bear, *Ursus arctos isabellinus*; blue bear, *Ursus arctos pruinosus*; and Asiatic black bear, *Ursus thibetanus*. Clearly, the sloth bear, the red bear, and the blue bear are not tree bears. Where Nepalis report two bears called the bhui balu and the rukh balu, the literature reports only one, *Ursus thibetanus*. And no description of this bear fits the tree bear, especially the bear's behaviour.'

Shah Dev leans forward, his eyes sparkling. 'Let me be clear. Are you saying that science does not know something that Nepali villagers know?'

'It appears to be.'

Shah Dev hasn't worked his pipe since the story began. Reaching for the phone he dials his brother Prince Gyanendra, who knows about wildlife conservation. With the instructions I just overheard, I expect that the prince is already on his way to the palace.

'What does this mean?' the king asks, turning to me after studying the ceiling in silence.

'It means we need more evidence. When we examined our first skull at the Smithsonian alongside the many *Ursus thibetanus* skulls in their collection, we found our skulls to be smaller and more delicate. But specimens of one bear don't prove anything except maybe that the Barun bears are small. So I returned to the Barun last week to purchase ground bear skulls, but the villagers claim that they do not have any of that type. I got two more tree bears but no ground bears. If skulls of both types can be collected from the same jungle and are shown to be different, that will prove that two different bears exist.'

'Collecting skulls will be easy. First we shall arrange the necessary permits for the endangered species, and then I will tell my
hunters to go and shoot both bears.'

'I already have the permits.'

Shah Dev settles back, remembers his pipe, scrapes the bowl,
and again lights up. Smoke curls. Over the years that I have
known him, as people keep coming to him with personal requests,
Shah Dev has consistently demonstrated awareness of how what
seem to be little things can link together unexpectedly to become
big. As he has been increasingly surprised, more caution has risen.
Smoke curls up. Monarchs plan across generations.

'We must be careful,' he says. 'If we do anything too quickly, we
will overlook something important. And if it turns out that you
are making a mistake, it should be clear that it is your mistake,
not ours.'

He is right, of course, and not just from a royal perspective.
Science is science because its perspective is also careful—seeking
to explain the Yeti I've tried to be careful, but this man is the king
of a nation whose mascot is the legend I seek to demystify. The
tree bear is what we have spoken of, but I need to alert him to
the fact that this bear idea, as it connects to the Yeti, may in time
be taking money and magic from his kingdom.

A side door opens. In walks Prince Gyanendra Bir Bikram Shah
Dev. I rise and step towards him. Walking past me, the prince bows,
and still holding his bow, continues to walk forward. His Majesty
remains seated as the other approaches him genuflected. The older
brother offers the younger his hand; obeisance is performed. This
deference must happen each day when they meet for the first time.
Only after that rite is over does the prince straighten. Firm of spine
and always polished in speech, he greets me regally.

The king brought the prince because he knows the latter thrives
on catching mistakes. The prince will pounce on any flaw in
my thinking.

'Dan'l, tell your story to my brother.'

'Your Royal Highness, His Majesty and I were discussing a discovery made by Bob Fleming, myself, and our families. We've conducted wildlife observations in the Barun Valley over the last year. The villagers there claim that two bears live in their jungles, the bhui balu and the rukh balu. The bhui balu fits the scientific descriptions of *Ursus thibetanus*, the Asiatic black bear, but the rukh balu seems unknown, for there is no scientific description about a small shy bear that lives in trees. This bear, known to Nepali villagers, is unknown to the world. As of now, we've collected three rukh-balu skulls, one set of paws, and intriguing data on nest-making. All verify villagers' reports, but disagree with the literature. To clarify what is true, we now need bhui-balu data from the same habitat, especially skulls, to compare with the tree-bear evidence.'

The prince breaks in. 'Let me understand. You say that no scientific distinction is currently made between the bhui balu and the rukh balu. However, our people consider these to be different animals. Are you saying that science does not recognize this?'

'Yes.'

'So how do we test what you assert?'

'Two issues need separation, Your Royal Highness. The first is the rukh balu/bhui balu question. Are these two different bears, and if so, how are they different? The second question is: might one of these bears be the Yeti?'

'Wait,' His Majesty breaks into the conversation. 'You did not mention that. What is the Yeti information?'

'As Your Majesty knows, there are two types of Yeti evidence. One is what the villagers report. Basically, nothing can be proved with these stories. The other evidence has scientific substance, footprints have been photographed and, in one instance, plaster casts were made. The most famous discovery is by Shipton in 1951. You must have seen that picture—a track almost 13 inches long, very clear with noticeable large toes that look human. The second set of credible photographs, as well as the plaster cast, was taken by Cronin and McNeely in 1972, a print just under 9 inches long.'

I pull photographs from file folders. 'Your Majesty, please examine this rukh-balu print. When my brother-in-law and I found this, we thought we had found the Yeti. However, as you can see, the photographs show nail marks. Although the nails point towards it being a bear and not a hominoid, the shape of the footprint does not match that of *Ursus thibetanus*. There appears to be a "thumb". Cronin and McNeely's Yeti prints are similar to these, but their print shows no nail marks.

'Shipton's sighting doesn't solve the problem either, because he has only two photographs as evidence—a close-up and a long-distance shot. And in their close-up, they presumably selected the most human-looking print. John Napier, a British curator of primatology, obtained the original negative of the Shipton photograph and discovered that the lower section of that negative had been edited out when the picture was first published. With that, we see part of a second print showing what could be nail marks. More importantly, Your Royal Highness, Napier shows how the print could have been made by an animal setting its hind paws into its front paw prints. Napier did not know about the rukh balu and possibly an undescribed bear. If we add what we now know of a rukh balu having a thumb-like digit, Napier's overprinting thesis for the Yeti becomes more convincing.'

'OK, your point about the Yeti is understood,' the prince says. 'However, you said that there are two issues: the bear and the Yeti. Tell us about the bear.'

'Well, the question is how the rukh balu and the bhui balu differ. Three explanations can be offered. First, the difference may be based in gender: the ground bear is the male, and the tree bear the female. A second explanation might be that the tree bear is a juvenile ground bear, a yearling that has left its mother but is still undersized. The third explanation is what the villagers maintain—that the two bears are different. Why not test whether the villagers are right?'

I continue, 'There may be an easy first action to sort all this out where we could use your royal help: Your Majesty might own a tree bear in your Royal Zoo. I examined this bear through the cage. The size is right and its front paws appear to have a thumb-like digit. One zoo bear by itself proves nothing, but it is of interest. Could Your Majesty arrange to have the animal tranquilized and let me examine it?'

The king touches a buzzer behind his desk. Before the buzzing stops, the door opens and the aide-de-camp is inside, standing at attention.

'Tell my private secretary to come. General Shushil—I want to see him, Narendra too. Tell Bishou to come.'

The king, the prince, and I now circle into a conversational holding pattern. His Royal Highness politely queries, 'You were in the villages recently?'

'Yes, I made a quick trip to the Barun to seek more bear skulls. While there I saw a new temple built by Your Majesty's government. The villagers appreciated it very much.'

'Which Barun villages?' asks His Royal Highness.

'Where the Barun and Arun rivers meet. This is an interesting place with unusually high …' I stop; neither brother is listening. People are filling the chairs along the wall. As they enter, each entrant bows at the waist, palms pressed above the forehead, and walks that way to a chair where he mutely sits, shoulders bent in deference; their faces show obvious consternation. All remain silent as we await one last arrival. Then His Majesty looks at his staff and says, 'This is my friend, Dan'l. I have known him for many years. He has some interesting discoveries. Please, your report, Dan'l. But only about bears.'

I explain about the tree and ground bears—that though Nepalis know both, Western science knows only the latter. 'If the villagers are right and there are two bears, this will be an important discovery for Nepal. I have asked His Majesty for help.'

'Yes, I want to answer this question. You are to help Dan'l. He wants to study the bear in the zoo. He wants to research in the Barun Valley next year. I promised the help of palace hunters. He is my friend and is careful, and he will not violate any of our national regulations. Also, he has the needed scientific collection permits. Make the arrangements.' He turns to me and says, 'Dan'l, I think this will be enough.'

Everyone stands, poised to leave. I rise too, but hold back while the others depart. His Majesty stands, looking at me with his hands clasped behind his back as I say, 'Your Majesty, thank you. It is always wonderful to see you. Thank you for your support.'

My friend grasps my hand, drawing out the moment by not letting go; a smile opens into a warm grin. 'Dan'l, it is going to be fun to watch you and this. I'm sure there will be more surprises. Do a good job with this. From what I know—and take this as suggestion, not command—the Barun's wilderness is more important than the bear. What we do for conservation in the Barun affects the adjoining areas, such as Everest and also the Arun Valley hydroelectric project. And we need to rethink our national park management. Let me point you to these issues as well, not just the bears. Seek conservation that affects the life of my people. We need ideas about how to be more effective at protecting what we have, especially ideas that make life better for my subjects.' He gives a gentle squeeze, and then releases my hand with a formal handshake. 'Good luck.'

At times after my meetings with Birendra Bir Bikram Shah Dev, especially when consideration had been of the whole well-being of his kingdom, I felt that this man was, in fact, blessed with divine insight. He held binding his country as one people to be the highest priority, and he looked into the future with insights few Nepalis had. He was very suspicious of Nepal's neighbours using his country for their national purposes. However, he was

11.2   The Author and His Son Jesse at the Zoo—the Bear is in the Net

Source: Robert L. Fleming

also human and perhaps overly cautious. Tragically, of course, he has been assassinated; the deed, I know for sure, done by his son frustrated with the old patterns of authority.

ॐ

THE NEXT MORNING AT 7:15, Bob Fleming, Jesse, and I rendezvous at the front gate of the Kathmandu Zoo. Seven government officials wait in the morning fog. They had orders from the palace asking them to show up. That's all they know.

'Our task is to study the black bear's feet,' I say.

They nod. A command from the palace to show up at the zoo early in the morning was strange enough; that the purpose was to study a black bear's feet made the command only implausibly stranger. The director calls for his tranquilizer blowpipe. Another among them calls for tea. After half-an-hour the gun arrives. Nepal's chief ecologist suggests a different tranquilizing drug. We wait for another hour and drink more tea. Finally, a dart is injected. The man who called for tea suggests that the plaster-of-Paris casts we want of hind paws overprinting forepaws will be more accurately made if first, a negative mould is made if the first imprint is in dung which has a smoother consistency than mud; then the dung mould would be used to make the plaster casting.

The bear weights 52 kilos (119 pounds); cusp wear indicates that it is middle-aged. It is a female and she indeed possesses something of a thumb—it's unclear how it would look splayed out in the snow, but it splays nicely in dung. All in all, a promising tree-bear suspect. But the series of casts we make is most intriguing. With slight changes, positioning the hind foot on top of the front foot, we have a bear making gorilla-like prints, and casts where with a different positioning of the hind paw on the forepaw we achieve prints with thumbs like ours on the Barun ridge. Also, longer cast prints are similar to Shipton's except for its puzzling wide middle toe.

The great 1951 footprint mystery now has its first hard explanation in sixty years since Shipton's photograph. A known animal is shown to be able to make the unknown prints. Using a tranquilized bear, with further moving of feet in the dung, it was possible to position hind feet on to forefeet such that not only the 1951 print but also the other Yeti 'mysteries' are explained.

Through the morning one man was particularly creative. It was his idea to order hot tea, his idea to cast the footprint in malleable cow dung to form a smooth mould. This talkative man of today was mute yesterday in the palace; Bishou Bikram Shah is a relative of the king, the deputy private secretary to His Majesty, the master of the royal hunt, and the keeper of the king's private lands.

A few evenings later, some of us who were at the zoo, plus three others, gather at Bishou's house. Among them are Hemanta Mishra, the chief ecologist for the Department of National Parks; Tirtha Bahadur Shrestha, Nepal's leading botanist; Rabi Bista, a hotshot forester; and Kazi, a bird expert whose field time exceeds that of any Nepali scientist. Looking around the room I realize that here are the kingdom's most informed jungle experts, plus each has had a little to drink. Casually, I ask, 'What do Nepali villagers think the Yeti is? Not what you think, but what do the villagers think?'

'Dan'l, the Yeti is our mascot the way the bald eagle is America's,' says Hemanta. 'Nepal is a land of Mount Everest, the tall snows— these things which no other country has make our small country well known. But Nepal is also a land of mysteries, of Shangri-La, and it is for these that westerners come. The biggest mystery is the Yeti. If Nepal explains what is unknown about the Yeti, we lose a lot of our magic. When I fundraise for tigers, rhinos, or pristine jungles, the Yeti is my ally. I don't mention its name, but it's on everyone's mind, the most endangered animal of all. If the interest in my speech is dying, Nepal's wildness is immediately strengthened with a passing joke so that they do not think I truly believe in it. A joke confirms that Nepal's other animals must also be rare.'

From one end of the sofa, Tirtha Shrestha breaks in: 'But Dan'l's question is not whether Nepal needs the Yeti. His question is about what Nepalis think of the Yeti?'

Rabi jumps in, 'In my experience, it depends on the ethnic group you ask. Firstly, in the Nepali language we didn't have a word for the Yeti until maybe ten years ago. Yeti is a Sherpa word. We didn't directly bring this word to our language, but as Mishraji said, we brought 'Yeti' into Nepali from English. Dan'l, if you ask a Nepali about the Yeti, a different answer will come depending on whom you ask. Villagers may not even know the word.'

Tirtha waits politely, not permitting his argument to be side-tracked. 'While helping write the fourteen volumes of *From Mechi to Mahakali* that describe all of Nepal, I talked with many people, and my thought coincides with Rabi's. On the Yeti, different Nepalis have different thoughts. But there is one belief that is common to all. All Nepalis believe our jungles contain the bun manchi, a jungle man. Is the bun manchi the same as the Yeti? The answer is no. I have asked for descriptions of the bun manchi and the Yeti, and the descriptions differ. But consistent across Nepal is the idea of wild men living in the forests.'

'The idea is strong,' Bishou acknowledges. 'In fact, sometimes I believe in the bun manchi myself. But when I ask hunters to show me the bun manchi, I get nothing. Never have I met anyone who has seen the bun manchi. Villagers only see their crops damaged by the bun manchi. For this damage an explanation is needed ... and often they ask His Majesty for compensation.'

'Maybe the bun manchi is like our Hindu gods,' Rabi suggests. 'I cannot describe our gods in English because your language lacks the needed ideas. Hindu gods live in another world, spirits that have moved into places around us. We have this idea that in an idol it is possible to be real and not just a symbol, but real in a nonphysical way. The idol is God; yet the gods are not idols. Ideas overlay; they are different faces of one god. Maybe the Yeti is part of that world that overlays into this world. In your English insistence on logic, you cannot accept different real realities—physical and historical and spiritual and something else—among which there is a disagreement even though we see consistency. I need more beer! I didn't say it right, but I am right.' Rabi fills his tankard from an open one-litre bottle on the coffee table.

'I understand,' I say to Rabi in Nepali. 'English talks about non-physical realities, but does so as though reality is a physical prod-uct, that reality can move between the physical and nonphysical. When I get off the aeroplane while returning to Nepal and start using Nepali, not just my words but my feelings also change. The

Nepali language has room for every idea, like a bus has room for whatever people bring—animals or baggage or new people. People who know only English are seldom aware of how product-oriented their ideas are because of the nature of the language in which they are conceiving those ideas.'

Rabi claps. 'Nepali as an overcrowded bus! Good, Dan'l, very good.' Hemanta and Bishou laugh. But I know I use flimsy allegory.

Having politely waited, Tirtha resumes, 'Your question was what Nepalis think. I say that Nepalis agree that there is a bun manchi, but the bun manchi they describe differs from one valley to another.' Looking directly at me, Tirtha says, 'Remember, everything is shaped by caste. Our system is built on each position always having a lower caste. Yeti, like the bun manchi, dignifies the villager. For each caste, it is more important to have lower castes than higher. For a villager on the hillside who is struggling with people with bigger fields, having the bun manchi raises their otherwise low status. It makes a man feel more civilized. With the bun manchi, even though it is a wild creature it is also a human of sorts, and so that man in a caste society is no longer at the bottom.'

Tirtha continues, 'Who are these low-caste people? They have land near the jungle; though the land is being cleared, it remains a jungle. Maybe, technically, they are Brahmins, but they are poor and live on bad soil. They want to be civilized. The bun manchi makes them more civilized than that jungle person they talk about.'

'Very good, doctor sahib, very good.' Hemanta has been listening intently. 'Maybe such poor villagers also find the bun manchi convenient to explain why their fields are so often eaten, because they are poor and cannot afford to make fencing. The bun manchi gives them an excuse for being poor.'

'Ah, Hemantaji, you keep coming up with the economic explanation,' I say, having learnt years ago that as a white person I must never comment on caste. 'Maybe Hemanta is wildlife's chief economist as well as chief ecologist!'

Bishou roars with laughter. 'Very good, very good. Wildlife's chief economist.'

'A good thing, too,' I add, 'for Hemanta has raised a lot of money and saved a lot of Nepali wildlife with it.'

'I have another point,' Tirtha continues, looking at me. 'I said that Yetis make the villagers appear more cultured. A part of not being an animal means having stories to explain who we are. I give an example. Americans characterize Russians as "bears"; doing this makes you "not-bears". You spend billions of dollars projecting the Russians as bears because of your belief, while the rest of us believe the Soviets will not attack. Maybe the Russians are dangerous bears or maybe not, but there is no doubt that you have grown a mythology that favours how you want to see yourselves.

'Likewise, Nepalis have mythologies to explain who we are. If I convince villagers that there is no bun manchi, I take away some of their self-identity. You just heard Rabi bhai—Nepalis have answered the Yeti question, also the bun manchi question, even the tree bear/ground bear question. It is you that made the Yeti into a mystery; you who are trying to fit it with science. The mystery would be answered if you accepted our context.'

Bishou sets down his beer. 'Good, Tirthaji. The bun manchi to us is an answer, and to the westerners, a question. Let us go to other room. I have some special meats. You must guess what they are.'

'Give us a hint, Raja,' I ask. 'Which jungle do these meats come from?'

'That is your mystery, Dan'l. All I say is that there is no bear meat on my table. Please come.' Vijaya, his lovely wife who is skilled at cooking the exotic animals of the jungle, opens the doors leading into the dining room. At the table are two types of yogurt and a platter piled high with garnished rice. I count five meat dishes. Two are fowl and one, I bet, is partridge. Two others are red meat: the greenish grey may be wild boar and maybe the other is sambar stag. Is that fifth one the tiny barking deer?

In the midst of Nepal's problems, as human population pressure reduces animal habitat, wildlife conservation has been very successful. The diversity of meat on this table confirms recent data according to which in areas under management, there is almost a surplus of wild animals. (Seven per cent of the land had been protected then in 1984; with additional land protected by 2016, now a quarter of the country is conserved.)

ᘒ

WHAT WOULD THE YETI'S LANGUAGE BE? In searching for footprints and exploring legends, explorers sought an animal. As this search continued, local people connected increasingly to an outsider idea about the Yeti rather than their knowledge. The outsiders assumed that they pursued a reality that was locally known—their error was in not understanding the original meaning. What followed then is shown by the quest overlooking the role of language. And the idea of discovering this wild man can be explored across many language landscapes.

First, consider *if* the Yeti exists as a true wild hominoid—it must have a language. So what would the Yeti's language be? Moreover, if a wild hominoid has succeeded in staying secluded despite a century-long search, it must have a sophisticated language. Such supreme crypto-locational seclusion would be impossible unless the language allowed it to understand people and communicate among themselves; the animal moves and eludes as explorers and villagers keep coming. So a real Yeti not only has its own language, it probably also understands ours, at least Nepali and Sherpa languages, maybe even more languages. This is a lot more than guttural grunts and primal screams.

Moreover, its language will equip it to 'read the jungle' with accuracy. For, in order to hide in a diminishing habitat, it must be using that habitat with sophistication not only for hiding but also as its home. The awe-inspiring abilities of Lendoop would be child's play for such an animal—and the creature would need to share these abilities among its kind, hence requiring developed linguistic skills. It would not only be a genetic 'missing link' between humans and their ancestors; it would be a polyglot professor.

Thus, should a true hominoid be discovered, apart from zoologists and anthropologists, the linguist community would be euphoric. Languages that function in jungle dynamics would be understood beyond the frames of words, hand signs, tones, and possessions. Such a hominoid might open a Rosetta Stone-like path so that humans could understand the wild. If the Sami people of the Nordic lands have about 200 words to describe snow, the linguistic skills used by the Yeti might be broader.

A Spanish speaker makes every noun carry a masculine or feminine character imbuing sexuality into life. Indians and Nepalis locate verbs last, causing their sentences to end with action. We know, for example, that in languages using touch, new worlds open beyond words. Understanding by touch is much more than the opportunity for a Braille speaker. For think of the understanding that comes to a cat that navigates by pressure on its whiskers to food, or a snake whose ears do not hear but feels sound through the ground. Smell gives a language for ants that assemble a lexicon from four simple scents. Dance also speaks where bees tell each other of pollen-rich flowers and starling flocks communicate by wingtips. Ponder, then, what language levels a Yeti who has been so magnificently hiding would offer.

Language maturation has been underway since the beginning of genetic evolution, not only for hominoids but for all life,. Multidimensionality of language is a facet of all life forms as is the process of living. We know a little about this dimensionality of language in the worlds beyond humans (for example, between whales or between elephants), knowing that these beings speak in the present individual-to-individual and across generations carrying messages from individuals then dead but speaking to individuals now living. Languages are being continually developed. Watching a string of ants communicate as they move is clear, but what is it like in comprehending the world to live in a language framed by mere pheromones?

To believe that human language is supremely sophisticated, or that one of our languages is more so than others, is potentially proof of ignorance. It also overlooks an important fact: each life form is continually evolving its language to fit its evolving circumstance. Languages, to state the obvious, speak from the scope of their experience. The incentive to evolve language for all beings is a requirement of life, for if a language is made more capable, its users experience a rising quality of life.[1]

But then, at some point, languages also die. For the Yeti, I wonder how its language would be dying as its habitat so evidently diminishes. As languages die, more collapses than just words and syntax. In this extirpation, the paths to understand life are disappearing. Language withers before the physical species becomes extinct. A withering language will parallel the deflating of that species as the relationship-building function of language declines faster than the decline in numbers, driving lost opportunities for communication. The separation from use of one's language is akin to an individual that may remain locked like an animal in a zoo, where excluded from the exuberance of once-enjoyed appreciation, the dynamic disappears that caused language to bind its members into a whole.

Language is, of course, larger than words. A language is not defined by its components but by *what it accomplishes*. So language mutation advances as life circumstances change. This is shown by the evolution of the phrase used in Nepal to greet someone. In 1961, on meeting a person we would say, '*Kahanh jane ho?*' (Where are you going?) But as Nepali interactions changed from the greeting exchanged on rural trails to urban connection, the word used now is 'Namaste', an expression that entered Nepal from outside its mountains, drawing on the Sanskrit '*namah*' (I bow) and '*te*' (to you). Not only did language expand geographically (like the word 'Yeti' going from Sherpa to English and then Nepali) but it also began to include action with words (a slight physical bow and palms pressed together).[2]

For humans, the great language enabler came when brains got larger, a change written in the fossil records. Simplistic steps of communication had been progressing. But when brain capacity expanded, new types of language grew, because on to the fossil record was documented the historical record shown in artefacts and art. As collective expression grew across generations, cultivated by intra-species communication, a further expansion of evidence mounted: civilization.[3] A collective brain such as the hive-mind that is civilization that talks by behaviours speaks perhaps most authentically as aggregated actions.

Thus produced is the ability to create groups and function as groups. The function of language is connection. Languages of behaviours gather this process. While mutual protection and sex are other relationship-creating functions, the multidimensionality of language is the primary gathering frame of civilization. (Sex, it must be said, does not appear to be growing more sophisticated through the ages, and methods of protection appear to be growing mostly in technology.) Through homo sapiens' use of language we reshape the world from what we knew to that which is new.

Language growth for each individual is striking. Language grows through one's life. While physical ability decreases after one-third of most beings' lives, language abilities mature until the narrowing years before our physical death. Elephants grow in wisdom.[4] Grandpa observed older tigers and leopards that he termed 'cagey'. Wisdom is accrued to almost all life as it ages, as understanding layers with complexity; the connecting matrix that brings this together being language. Such language growth—of expression of our behaviours in engaging the world—functions on different planes from the biological, while profoundly intersecting.

Even the word 'Yeti' has changed its meaning. A remote Sherpa word from isolated glaciers now animates a wild possibility for humans. In this, footprints across wild snows became like punctuation marks, giving emphasis, but the ideas laid down were extensions of human desire from our early DNA. The message brought through this language of connection moved out of the remote Himalaya to new life in London, Hollywood, Bollywood, and the markets of Kathmandu. For in our brave new world the definition of real does not require real existence. It requires the ability to present a reality on a glass screen and in the imagination; that is, reality becomes real in the world of human manufacture rather than in the real world.

The Yeti became a projection of an inside reality. It did this through a language of desire. The human experience in London, Hollywood, Bollywood, and Kathmandu developed a means to speak from inside homo sapiens about a relationship with the wild that was otherwise lost from a world where those people in London, Hollywood, and Bollywood had long been separated. Language enabled reaching out from that inside, creating expression. If not by language, how else do we bring that which is inside us out into an engagement with others?

We may talk of objectivity, but all beings always view the world from inside out—though perhaps trying otherwise. And to enable that extension what we have created is language. And as we do so, we are losing the old. The Yeti (and other creations of the glass screen) has brought to humans a new life from the wild.

If we understand languages this way, expressions of life to life are therefore everywhere. In a naturally wild world life is 'talking' to each other, growing a conversation of living—being matured is a civilization not of human making but made of Life's multiple selves. Outside the glass screens, life is communicating in multi-majestic ways. This is visible, for example, in what a blossom tells us about the earth's position as the flower changes its form with the passing day. The sun communicating to the flower is similar to an actor communicating to a film-goer. Messaging happens on multiple levels, many of which are overlooked, as the blossom is communicating to insects called by the flower to feed, and then allowing the flower to communicate across generations of life itself to new generations through heredity. That flower which is crossing life generations also is communicating by turning people's behaviours, causing, for example, people to be affectionate, procreate, and generate new generations of humans ... who may, then, consume life in order to live their lives.

When the wild is the context, all life forms are communicating. This is a feature of being alive. And if a hominoid Yeti does not exist, then its non-real existence is evidence of an astounding further life feature: non-living animation communicating with the living. In following the footprints, I discovered that what created the Yeti was something more than those footprints. Language was working in many ways. Emerging from a desire for a wild hominoid, a creature went out from those footprints across the peopled planet, indeed binding people together and also binding them to a wild being lost. Language allowed the non-real to become real.

Notes

1. Michael Tomasello, *Origin of Human Communication* (Cambridge, MA: MIT University Press, 2008); David Crystal, *The Cambridge Encyclopedia of Language* (Cambridge: Cambridge University Press, 1997).
2. Sometimes the word used is namaskar, which literally means 'I bow to your form'. Here again the palms are pressed and there is a slight genuflection—speech originating through a language of body movement, not words. In English, we seem to be tiring of our once almost-universal 'hello', and, as the language evolves, we are turning to 'what's up?/good-to-see-you' or other experimental salutations. Perhaps our earlier greeting (hello) indeed needs to change, having

lost its origins in old German, according to which 'Halo-ing' means 'fetching' a ferry, or in the cry of 'Hollos' that originated from old English hunts.

3. Terrence Deacon, *The Symbolic Species: The Co-evolution of Language and the Brain* (New York: W.W. Norton & Company, 1997); Tecumseh Fitch, *The Evolution of Language* (Cambridge: Cambridge University Press, 2010).

4. Carl Safina, *Beyond Words: What Animals Think and Feel* (New York: Henry Holt, 2015).

twelve

Back in the Barun

It is 18 November 1984. After nearly two years I've returned to Makalu Jungli Hot'l. Those who accompany me seek to unravel the enigma of the tree bear. But for me the objective of the search is changing—the question to be answered goes beyond a bear or a valley, turning to questions about how to preserve the wildness in our lives. From this pristine valley sheltered by the highest of mountains can flow a lesson of high importance.

Near the campfire scar, rusted tin lids peek through from under leaves. Around our former garbage hole is junk once thought buried, dug up by smell-curious jungle denizens. A slivered plastic circle that earlier wrapped pepperoni gilds a bamboo shoot—items cast aside thinking that the wild would consume them, detritus from our ways of living. Like an animal, I prowl checking corner-post markings. We came looking for a wild man, and to make that easier brought the non-wild with us. In leaving, we could have returned production's debris to 'civilization' to leave the wild 'pure'. Would that have made this place 'wild' again? In contrast to our single visit, shepherds who have camped here several times left a flocked-trimmed meadow that made the wildness welcoming.

The air, 4 miles above this camp, grows increasingly populated. More than 4,000 climbers have topped Everest; hundreds
of thousands visit her base which itself is 2 miles above, and has
half the atmospheric pressure of this wild jungle pocket. Such
intense inhabitation of the once-wild is just that which is on
Everest where dozens of dead also now permanently dwell. Flying
daily beside this emblem of human conquest of the wild are jets
carrying hundreds of people, each thickening the air. But in this
secreted valley pocket is natural balance. In this saved remnant
I have an opportunity.

Some go to the wild to conquer; others to escape. But whatever
the purpose, human impact on the once-wild is clear. Imprints now
are no longer of wild animals—where 'Nature Mama Rules'—now
Nature responds. We may not like the response—jackals and bears
enter our cities from the wild where Nature Mama once ruled—
but Nature now responds to us.

My opportunity is to make an impact. One person's role may
seem insignificant, as throwing away a sliver of plastic may seem
not to make much difference. It may seem the sliver would be lay-
ered under the leaves. But a bamboo shoot has come up inside it,
lifting that one thin wisp of plastic high as a flag. Nature responds
to our every single action. On returning to Makalu Jungli Hot'l,
I cannot go back to the old wild—it might seem this place is
at the end of the earth—but here too our actions of change can
be lifted high.

Life grows forward, always. This understanding is pivotal. From
the present, a new future is being created. Nature is like time,
going in one direction, forward, never returning to what it has
been. This means that Nature is always losing its prior balances
with a new always coming, and an impossibility always with us
to regain the old. On this understanding, a fallacy in the old
thinking is revealed. A protected pocket may be re-established
around a natural remnant, but that is not Nature's; it is human
manipulation. National parks have been created to ensure that

such pockets remain. But what they try to encapsulate has been lost, life balances now only preserved in these time capsules of the past.

But I've not returned to create another protected pocket. I have come to embrace participation with the wild, recognizing the power of people in the global ecosystem. There is a dynamic to be used in the fact that Nature is responding to us. What is envisioned in our coming here is more than the Makalu–Barun National Park (although that has been emphatically agreed upon). What is intended is more than the well-being of Nepal (though the king wants that). We shall start here an idea that will grow. Moving from the concept of protected pockets, or when that is even lost and Nature becomes specimens on sliding trays in museums or rare animals barred apart in zoos, this new vision seeks to embrace Nature with people-participating protection.

My quest is to prevent wildness from being fenced-in—where instead of looking through the fence, or flipping through glass screen channels, or pulling out skulls of the dead, people learn to live with the wild, a continually learning engagement with life. The first part of this approach is to release the compulsion of Man's dominion over Nature, which is the out-of-date view. (Even when control is possible inside protected areas, threats more significant lurk outside.) Partnerships will be engaged in this new approach, from the community to the national and the international; in people must be the preservation of the wild, if, as is the modern evidence, they are the danger.

The study of Sanskrit gives a metaphor of where the process must go. In its origin, the word 'jungle', which usually prompts thought of tigers and bears (that many seek to protect), was used to describe domesticated fields that returned to brambles. This earlier word use meant a once-peopled place going wild—not the wild place now threatened by people. We have lost that meaning (the domesticated returning to the wild). Can we flip today's meaning of protection back to the original Sanskrit? To do so requires relinquishing the premise of control, a premise deeply ingrained in the modern way of life.

Sanskrit gives another metaphor. Growing a new jungle comes iteratively; in *iteratively* is the insight into 'the how'. Iteration comes from *itara* or 'other' in Sanskrit—the method involves 'turning to the other', not randomly but by experimenting with alternatives. In a present age that lacks knowledge of the end point, we have the process to get there. We shall learn the way. Accordingly, we have come here now with a group of scientists to understand from the Barun. This trip is the beginning of experimenting with ideas; it will take us forward in many years of iterative growth.

To do so, another flop within a flip is needed, a change in perspective. Ecosystems were once defined as systems of Nature, such as ridges that defined watersheds. Protection meant 'managing' ecosystems. That is a narrow perspective. Biology is no longer (if it ever was) just flora and fauna. For example, economics must be engaged (and much more than markets for bear gall). The growing populations of endlessly aspiring peoples must be dealt with. Nepali and world politics must be recognized. And the world is awash with information—seemingly, all knowledge is available and it is not clear what is true. These dynamics are of human making. Their collective is now what defines the ecosystem (expanded from the natural sciences) that will protect or destroy Nature.

So answers must grow from encompassing scientists, governments, and people around the world. A shocking change follows then: instead of viewing protection in a natural ecosystem perspective, the new definition moves to 'operating systems', shifting from Nature's coding to human. Even genes are not just coded but also adjusted—with this comes great danger. While it happens, we risk slipping into the conceit of playing God. What causes operating systems to work is not engineering the answer (prescience

12.2   Rhododendrons and Trees in the Barun Valley Dressed with *Usnea* Moss

Source: Author

or omniscience) but setting up more rapid participation, that is, iterative processes leading to answers driven by response—driven by using principles. Operating systems are effective because they allow us to mimic change of continual response, evolving answers that fit. The answer is not known beforehand. The answer grows through experiments.

Reality is not being created by this process. We shall never master the universe where we live, move, and have our being. Trying to do so attempts creation out of proud thoughts and vain desires. But to position our place, going beyond the blindness of our hearts is what will help develop a refuge. Otherwise we slide into the curse of Ezekiel: 'Is it not enough for you to drink the clean water? Must you also churn up the mud with your feet?'[1] Or as Jeremiah said, 'I brought you into a fruitful land ... but you defiled it and made the home I gave you loathsome.'[2] In returning now, I have returned on a new quest. From this jungle where a full earth still soars to heaven, we have come to open an opportunity.

෧ා

I START SETTING UP OUR HOMES OF ALUMINIUM and nylon while Lendoop leads John, Derek, Bob, Tirtha, and Kazi into the jungle. Then, three hours still left before supper, I push through the stinging nettle on the edge of the camp. After an hour I catch up with the group. A palace hunter carries one of His Majesty's glistening. 30-06s (royal bear protection). As the team climbs, John and Derek Craighead are picking up plants, constantly commenting. Bob and Kazi walk apart discussing the birds they hear. Lendoop glances right and left as his kukri swings. As I trailed this party I noticed that Lendoop's been following precisely the trail of twenty-one months ago that led to the ridge with our footprints in the snow. How does Lendoop see the old signs? The earlier trail was nothing special; a climb straight uphill. But Lendoop, I suspect, plays a woodsman's game, retracing the old trail for the challenge of doing so.

The warm temperate zone we've climbed out of is a place of rapid decay. My earlier measurements were confirmed on the trek this time as we came in; the 1-foot deep loam on these slopes is the deepest recorded for the Himalaya. As Nepal's population grows, people will start cutting back the jungle. Until now, without people removing the litter and bushes for fire and fodder, slopes grow the decay of life, and that is holding water for the dry season in the valleys below—and with that water, a wildness without people nourishes the growing of food for people who live valleys away.

Bob and John are exploring under an oak, looking at bear scratches on bark and broken acorns. As they do that, Lendoop points up the slope, and he and I walk towards the tree. Out of the fork of a sapling, he pulls out a square-shaped faded candy-bar wrapper that Nick and I had placed as a trail indicator. The paper is stained almost like the colour of the tree bark, but Lendoop spotted it because its shape was different from that of a tree-bark flake.

Climbing higher, Bob hears the white-gorgeted flycatcher. Three quick notes, and he recognizes a voice he's heard but four times. There are 835 recorded bird species in Nepal, more than those in the United States and Canada. Bob has seen virtually all of them—he wrote the book *Birds of Nepal*—but modesty prevents him from telling me the precise number. He knows thousands of birdcalls worldwide; again his modesty means we'll never know exactly how many. One of those calls is the three quick trills we just heard. Earlier, as we passed through the lower jungle, he stopped mid-conversation on hearing the spotted wren-babbler. It called once. Three years ago, he had heard this call once in Sikkim. Now he's off with the white-gorgeted flycatcher, with Kazi on his heels. Working through the undergrowth from two sides, helped by another call, they find their bird, the 836th species for Nepal.

The Craigheads are seeing a cross section of this jungle, so different from the American West and Alaska where they've become leading authorities on the grizzly bear. It was good Barry Bishop pushed me. John and his brother Frank pioneered the tranquilizing of bears in the 1960s, then the use of radio collars for tracking wild animals, which is now a common zoological practice. In the 1970s, John developed methods for using satellites to map wildlife habitat, and now John and Derek are demonstrating how to describe ecosystem parameters using satellite remote sensing coupled with meticulous ground-based fieldwork.

Reaching an altitude of 10,000 feet, we stop; a good afternoon. Soon dinner will await us in the camp. As the group heads down, I remain under a birch that towers up through the canopy reaching higher than any other tree. After a while the birds change their singing. Other sounds become more regular. Sometimes, it seems, we know a lot, for in Latin names what we see becomes organized. At other times it feels as if we're losing what we once knew. The Latin organization is based in ancient genetic histories, and while biology may be about the study of life, life is about being alive. In singing their songs birds are not ignorant of relationships; some songs protect their places while others call potential mates to these places.

Thoreau's claim: 'In Wildness is the Preservation of the World' points to preservation coming from going forward. Had Thoreau said, 'In wilderness is the preservation of the world,' it would point to the past, the world as it functioned before people changed everything. As all now respond to human actions, a new world is created. Nonetheless, 'Nature Mama' still rules as her human child changes the rules with the new world evolving as with a mother raising a child towards new-fitting creation.

On this expedition I am a planner at work among scientists. Success in planning happens when the planner focuses on the process. That I am also a white man in a land of browns is a parallel starting point. Because I lack citizenship and do not pay taxes to this land, what I bring are ideas. To the emerging process, my role revolves around participation in this jungle place and national space. Those who have authority own the place, and those who are scientists can guide the emerging direction. I have the opportunity to bring all—the king, villagers, government servants, and international experts—into a partnership, thus driving the process forward.

Albert Schweitzer, the white doctor who left the concert halls where he played J.S. Bach to go to the jungles of Africa, opened intricacies in the Sorbonne differentiating between the historical and eschatological Jesus; this polymath focused on balance. Civilization is an ethic, he argued. Seeing the gathering of people as an ethic is an insight that was different from the definition of civilization being made of cities and material production; it went beyond defining civilization as language or labour. An ethic of civilization defined people in a world shaped by their learning, discovering patterns through life lived like lines of music dancing in discipline, defining not by notes but how the parts interlace.

Schweitzer had the courage to play his dance with both music and Jesus, a dance of passion and compassion. His life instrument became the practice of medical healing, attempting to bring into concord both mythos and logos. In his going to this dance he took an organ piano encased in zinc. And when he arrived in the new-for-him world with his instrument from the other world, he found the wild already dancing with a civilization that had been growing for thousands of years. He had to let go of control. But as I press on towards my goal of sustainable dancing with the natural world, my mission is not to bring help—certainly not technology, whether encased in zinc or economic development. I intend to encourage ownership by people of what they already have—a process for people to own their futures.

A twig snaps behind me. I turn slowly. Has a curious mountain chamois approached? No, it is Lendoop, squatting amid the bamboo. I did not know that he had stayed. It's been maybe twenty minutes. Did he snap that twig as a gentle reminder, perhaps eager to descend to Pasang's cooking? For a village hunter, coming to the jungle with scientists offers sylvan opulence.

But this is the first time on this trip that we are together alone. From my backpack I pull out the snapshot of his family, taken on the porch of his home at the time of my skull-hunting race-in. On his hand-hewn porch squats his pretty daughter. The photos Bob and I brought are the first colour photographs of themselves that the Shyakshila villagers have ever seen. Lendoop quietly studies the picture of his daughter and home. Then, just as quietly, his massive shoulders start to heave. He is weeping. I wait, letting a tincture of time mend his pain. After some time, he tells the story.

'Three months ago, after your short visit, my daughter went with three other girls to graze goats above the Barun River. One girl, being thirsty, scrambled across the rocks to the river. My daughter accompanied her, for the river is fast there. As the first girl leaned from the rock and scooped water, my daughter held her dress.

'Suddenly one of them slipped—the other girls who were watching them from above couldn't tell whether it was my daughter or the other girl. The rocks there are always wet with the spray. Both girls slid into the white foam, heads not even popping up once from the water's tumbling teeth.

'One afternoon a week later, my wife learnt from women on the trail that a body lay on the riverbank downstream between the Num and Hedangna villages. I walked all night to reach that place. I found bones, only bones, flesh eaten by jackals. And yes, I thought them to be the bones of my ten-year-old daughter. And so, with a blessing as well as a gift to the gods, I sent her bones into the Arun's current.'

I sat quietly, waiting.

'Doing that caused many problems. Across from our Shyakshila is the Shibrung village. One week before I found the bones, a thirteen-year-old boy from Shibrung went through everyone's fields. Finding ears of corn missed by the harvesters, he collected these in a basket. Looking to make some rupees, he carried the corn to Khandbari Bazaar, which was three days' walk away from Shibrung.

'After two weeks of absence, his father set off to search for his son. Above the gorge between the Num and Hedangna villages, the father found the boy's empty basket behind some bushes. He asked around. No one had seen the boy. But they told him of the bones I had sent into the river. The father accused me of killing his son and stealing the corn. The police arrived in Shyakshila, and I was taken to Khandbari in chains. After spending thirteen days in jail, I was finally cleared. But now I owe 4,700 rupees as legal fees and fees to the judge. I have never had that much money, so the money lender in Khandbari will come next year and take my fields.'

Lendoop paused, but his story started again. 'Two weeks after coming home, my cow slipped and fell off a cliff. I borrowed another 1,000 rupees from the moneylender to buy a calf so that my family could feed on milk. I have sold my gun to make the first payment to the moneylender, so now I must earn my money from farming.'

For Lendoop the spiral of the Third-World indebtedness had begun—debts people like him would never be able to repay. Lendoop, who was one of the wealthier men of Shyakshila six months earlier, might now be indentured for life. Without his gun

12.3   A Now-Older Lendoop in a Tree He Spent the Night in

Source: Author

that allowed him to use his jungle skills, his remaining option was trapping animals. Would it be bears for their gall? Musk deer for their scent? But maybe high wages and a generous tip from our expedition might save him (which I later would give).

He looks deeply at his daughter in the photograph. I reach across to brush off tears from the picture's surface, and then slip the photo into his shirt pocket. A few minutes later, he and I rise, and like Nepali boys at play; we lope the slope to reach our camp. The scientists have eaten and are writing notes in their tents or under trees. Pasang has saved some supper for us.

∽

THE DAY IS 23 NOVEMBER 1984. My watch alarm chirps at five o'clock; it is still an-hour-and-a-half yet before daylight. Today our team will climb the ridge on the southern side of the Barun Valley. Two years earlier, we failed in our three attempts because of waist-deep snow. So we never 'binocular walked' the warmer south-facing slopes for 'the something' hoped to be caught exposing itself in the sun. In the shaded snows we found tracks, nests, and stories. This morning I shall climb up to that ridge. Yetis are always found on pathways where their tracks show them crossing one valley to another.

For a few minutes, I listen as Pasang brings up the fire on which breakfast will be made. Lying in the dark my mind starts the climb. Any ecosystem is complex, but here in the Himalaya, with life layers stacked one on the other, our earlier visit had shown us how complex these dynamics are. Now satellite radiographic images and high-altitude photographs show how the valley runs, where deciduous trees become coniferous, where conifers turn to alpine meadows.

But the satellite images lack the 'feel'. It is feel that gives this valley its huge sense of the wild. The feel can be touched when one lies in the dark, listening and smelling and filling in oneself with imagination. In darkness, understanding probes where sight cannot.

Last week in the lower jungle I climbed to a lower crest of this ridge. I couldn't see out even on that crest for the jungle was so dense. Climbing an oak, from a fork near its top I got a view of the mysterious Mangrwa Valley, whose drainage starts at a pass at the Tibetan border and tumbles from the north into the Barun. Lendoop reports big meadows there and a torrential waterfall. 'Imagine a river urinating from the middle of a cliff!' Indeed, an image to ponder over from the world's fifth-highest mountain.

As a spotted scops owl bids farewell with a few final calls to the night, I unzip the sleeping bag. Pasang and one helper sit by the fire. Aware that I prefer hot milk and honey, Pasang offers that. 'Pasang, I'll skip the drink. I need walking time now.'

He hands me five oranges, a pack of biscuits, and a stack of chapattis made the evening before. I smear peanut butter and orange marmalade on the chapattis, rolling each and stuffing them into a plastic bag. I apologize again for not taking his milk, feeling worse as I realize the insensitivity I've shown since he must have gotten up for me earlier than planned. Guided by my flashlight, I push through the nettles on the edge of the camp. While climbing, my mind stays with the two waiting for others to awaken so that they can do their job. I walked away from the camp to be alone. Colleagues, like family, are fires in the camps of our lives, and we deeply recognize the warmth they give. But the wild, especially this place where so few have walked before, is discovered by being alone.

Above the treetops of this jungle, the sky tinges grey. The night retreats, compressing towards tree trunks. In between the trees is where morning opens first. The trees remain like towers of darkness. As sight comes, the darkness is pushed back evermore. An awareness of distance expands along lines that could only be felt earlier. Maybe an hour still before darkness truly leaves, but as I climb now my world extends beyond the flashes of my light that allowed sight to pierce the dimensionless night. When night yawns to dawn, my beam clicks off.

As I crest where we found the 'Yeti' tracks, I find that the ridge is now snow-free. I see the rhododendron branches it walked on using them like snowshoes. Farther up, protected by shade, old snow covers the route, and I find serow tracks. With the snow, I will now be able to read which animals have been travelling this ridge. The serow, a short-haired, donkey-sized goat-antelope with white stockings is now rare across the Himalaya, for it is an animal of the dense jungle. This individual soon angled off. A couple of hundred feet higher, I see tracks of the lesser panda, a tiny red-and-white-faced cousin of the giant panda from the Chinese side. This individual was unexpectedly high this time of the year for bamboo shoots are not yet out.

Higher still, I come across tracks of a musk deer. Pressure on this animal from poaching is considerable. Maybe the way to save it, as proposed by Nepali biologist Sanat Dhungel, is to farm the animal and saturate the fragrance market with musk milked from farmed animals. That will lower the global price, bringing down the incentive for illicit harvesting. The Barun would be the ideal habitat for a musk-deer farm because in this high, cold temperate zone hangs the lichen *usnea*, one of the deer's favourite foods.

Domesticating the wild to protect the wild—is that the way to treat (or justify) this last-of-its-kind place? Soon the deer's tracks I follow are joined by a leopard's. Prey and predator's trails appear to be made yesterday evening while the snow was hardening. The two trails, the leopard following the deer, turn into a bamboo thicket.

Reading tracks in snow entails guesswork. The snow type, the wind, lower atmospheric pressure that promotes sublimation of snow, the canopy overhead, the sun and its angle, and the ambient temperature shape a footprint after the animal first imprints it. Looking at the leopard tracks I wonder again about Shipton's. That day he claimed to have picked 'the best one in the shade'. How did he define 'best'? Another question has long baffled me: what gave shade on a glacier? Did he mean a boulder, for it had to be high enough for something big to walk under it? Yet that does not make sense as few animals walk intentionally under overhangs; or, if they do, they usually go under it to rest. Shipton's mention of shade raises questions of credulity with regard to his report. Also questionable is the fact that he intentionally selected the most mysterious print, implying an abnormality.

From the high altitudes concerning this legend have come only mysteries. But usually biological relics endure at altitude, So if there was any physical reality beyond footprints in the snow, even just if only a little, such evidence should be found. Higher altitudes are devoid of moisture, bacteria, and fungi—nature's agents of physical decay—so for the animal that makes the footprints why do just the footprints endure and not any relics of the body that made them? On mountaintops in California's dry White Mountains, dead pine tree trunks have not rotted even after 2,000 years. The teeth of every other animal are found on Himalayan moraines, and yet no Yeti teeth have ever been found.

Why are footprints that melt and lead to empty trails the only thing found? Claims of Yeti skins have all reliably been attributed to known animals. Because no other physical proofs have ever been uncovered, the explanation that does not disappear is that the footprints are made by an animal, which, when its body parts are found, turns out to be a known animal.

❧

TRACKING, ALWAYS SOPHISTICATED, NOW IS AN ALTERED ART from the past. Move by every move, FedEx shipments are tracked. The stock market is tracked. Children are tracked through according to their abilities. We track the wheels of a car when driving, and our eyes track the open space between the print in a book. These are new tracking skills—and being lost is life's oldest, arguably, humanity's first science.

In all forms, tracking is logic. Stories by trackers are almost certainly humanity's first systematic narratives. Through tracking, our species found its way to food. Our species would not have found the food we needed if acts of gathering had been dependent on luck; from the art of tracking we then developed the science. From tracking we asserted logic over the land.

But humans lack animals' most powerful tracking device: smell. Each family of life manages fragrances differently, but the primate family scarcely manages with smell. Carnivores' trails reek with their concentrated urine and odours, for these animals worry little about being followed, protecting their food by fencing it with a noxious scent. And the feet of many animals smell, leaving odours imprinted in the land. Because humans' sense of smell is relatively weaker, we follow by sight. Footprints are viewed as shapes, but a lot more describes them from the sound made to the smell laid.

More than following specific marks, skilled trackers in using their sight read the pattern on the land as part of the land. A first vision is that of perspective. It involves not looking down but ahead, as far as the eye might detect what has been dragged or where brush might be bent back. Only then does the tracker look at the prints making the trail. The tracker is seeing these as roadways, ways of the quarry and ways of others who might also be following, connecting into continuous trails of animals gone before, assembling deeper meaning from the multiple tracks as well as direction.

While following specific tracks, a tracker walks beside them. Each animal has specific imprints that can be learnt. As people have distinctive fingerprints—tracks left on everything we touch—so does every animal, for hooves are nails on fingers. And except for snakes and birds, animal tracks made from four feet are available. And among these four one foot will likely have some abnormality: a chip off a hoof, a bend to a digit, or the manner one drags. (Snake belly trails also have distinctive features.) Spotting these is easiest when highlighted by a shadow that lifts ridges or slight indentations to view. Thus, the best time for tracking is morning or evening because that is when shadows lengthen these details, magnifying (as a lens magnifies fingerprints) what might be less easy to see when light shines straight down.

On hard ground where imprints do not reveal much, look for items unnaturally placed—a pebble turned, a broken leaf or twig. Disruption happens into and on top of the land. Though the foot-prints are found on the land, tracks also are left above. There may be signs in bushes or on trees or brushes on rocks that are made when they scrape them with their backs and necks.

Try to think like the quarry you follow. Human senses may be weak, but we can use our greatest gift, the mind. Explain the pattern when it is obvious and then extend it when not obvious. Follow the trail using the mind to predict. Knowing the habits of the quarry helps to get inside its mind, and it amplifies the trail. So while following one trail it is important to remember the earlier trails of the same kind. Was this individual looking for food? If so, where is its food in this land? Was it trying to get away? If so, where is the easiest escape path? As the animal is identified as a species, what also needs to be identified is the purpose that is making the path you follow. Is it headed to rest? If so, taking into consideration these habits, in what places does this species and age of animal feel secure? Was it en route to its secluded children? What type of home does this type choose? Is it headed for sex? It would be almost impossible to predict such a trail, but sometimes sounds made by the mating animals can help. When following by the mind, the spoor itself need be seen only occasionally.

A resting animal, its eyes being closed, usually lies with its nose and ears facing upwind so that its mind can sense danger coming from at least that direction while the eyes are closed, the animal is relying on the wind to carry approaching sounds and smells. In sleep, the senses of smell and hearing remain awake.

While other trails may brush against bushes, the trail of humanity rises above the land into the sky. When it comes to footprints, we leave a footprint no other animal can: carbon footprints. While other trails pass across the land, ours circles the planet. This larger trail took its energies from the life that died long back, and so this distinctive human trail not only circles the planet but also reaches back into time. And while the paths of others end where they are laid to rest, our enduring imprints shall disrupt life for generations to come. In my Yeti search, my mind often remembered the cartoon on my refrigerator—words a Yeti might say to a human if an encounter were to happen: 'Is it not strange people call me "Bigfoot", for yours of carbon is so much larger than mine?'

❧

IN THE AFTERNOON OUR GROUP SPREADS OUT ON the ridge where views extend to the Chinese border in the north and past Makalu to the Everest massif in the west.

Tirtha, though, points to the jungle. 'Before us is the wet Himalaya as it existed before people,' he says. 'You see in this valley seventy degrees of biological latitude—a biological view from Delhi to the North Pole, and it is all pristine. It is impossible to physically see that anywhere else now. Imagine going from Florida to the Arctic and skipping Washington DC, New York City, farms and suburbs, and the extensive human alteration of America's East Coast. But such a journey in a pristine habitat Asia can be done here. Here, a view of 5,000 miles on the flat is travelled with a turn of the head.

'Temperature change creates these habitats. For every 1,000 feet climbed, the mean temperature drops by three degrees. Today we climbed from subtropics up through temperate zones to reach this alpine zone. Above this zone is the arctic. This is just a habitat range created by the temperature. Add to that the way the sun's energy varies depending on ridge exposure; some folds of the land get direct sunlight while others get shade. Ridges also change air currents and with that rainfall. What is extraordinary about this valley is that all this diversity of habitat is a world never altered by people. The Barun may be a valley ... but it is also the whole Asian world.

'Everyone expects the Himalaya to have an alpine habitat, but it is much more. Where the Barun and Arun rivers join at 3,100 feet is a phenomenally low for Himalayan heartland, and from there with fringe specimens of the tropics, the habitats climb to 29,000 feet which is beyond the Arctic and on the fringe of Space. This valley's bottom begins in a forest that has *Castanopsis*, which is a type of beech, and *Schima*, which belongs to the tea family; as is common with the subtropics and tropics.'

'What's so special about the subtropics?' I ask.

'Today more focus is on the tropics because of their species diversity. So biologists prioritize this life zone. But to support human life, the subtropics are more important; we are animals of the subtropics, for this is the zone where we were nurtured as a species and where civilizations evolved. Look around the world. There is almost no pristine subtropical habitat left. In many places tropical ecology exists. But humans, after agriculture, almost completely changed the subtropics. Mesopotamia used to be biologically rich—it is where agriculture began and flourished—and now these lands are desert. China and India were lands of jungles and savannas, but today they are human food fields.'

'Well, what about the Barun?' I ask.

'In front of us is the pristine Asia, not just of Nepal but also of wet South Asia. What your eye sees there at the top is the Asian Arctic, then where we sit now is tundra Siberia, and going down below us are all ecozones to the subtropics—all wild. Tonight stand on this ridge and look across the valley—night makes it easier to imagine a wild world. You are on a ridge many might call "nowhere", but look out and make two journeys: one from the subtropics to the arctic, and the other back in time before

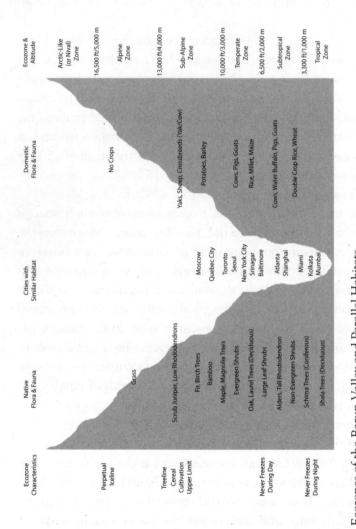

12.4 The Ecozones of the Barun Valley and Parallel Habitats

Source: Adapted from Tirtha Shrestha, *Development Ecology of the Arun River Basin in Nepal,* ICIMOD Senior Fellowship Series (Kathmandu: International Centre for Integrated Mountain Development, 1989) p. xx.

people populated the planet. You are not "nowhere" but sitting in middle of what Asia was, and experiencing life across time, and planetary diversity.'

๏

AS THE MOON RISES, I ROLL OUT my sleeping bag. On this ridge, a 'Yeti' walked by Cronin and McNeely's tents twelve years earlier. They had camped by a pond. Our site is almost identical, a few miles lower on the same ridge, and by a pond from which we've taken cooking water for the supper. If tonight a Yeti walks by, inside a tent is not where I want to be. So I tell the others as I unroll my sleeping bag: 'Tirtha told me to look out across the valley tonight.' But to be ready, I set a powerful spotlight near my head.

For Cronin and McNeely, the visitor probably came because it smelled food. Or did the Yeti come because it was headed to that pond? Water on a ridgetop? I hadn't thought about this—ridgetops in the Barun seem to be the only places with ponds. In this valley why aren't there any ponds on the mountain slopes?

About 20 feet downwind, I place a jar of peanut butter, lid unscrewed. I crumple a paper to make noise and tie that to my sleeping bag. I then connect to that open jar with a string, so that if something starts to take that jar the crumpled paper might awaken me but certainly will the pull on the string.

Ponds must be on these ridges because the sedimentary rock layers of the Himalaya lie horizontally, causing the water that seeps in higher on the ridge to slide along a shield to exit in ridge depressions, creating these pockets of water in the sky. A raindrop, pond drop, ocean drop—scale changes of water in planetary systems. I look at the stars contemplating my special space and time.

At 5:00 a.m. my alarm chirps again. Sitting up, a snow-dusted sleeping bag drawn round my shoulders, I sweep my surroundings with the powerful flashlight. No fire yet burns at the cook site. Not even mice prints show in the snow around the food packs.

Nothing has approached the peanut-butter jar. I look at the stars. We're at 15,100 feet, higher than any land in America's lower forty-eight states or in Europe. There is 40 per cent less air here than on the plains of India below. It seems I can almost touch the stars. But I have awoken to go even higher. Unzipping, a shock of cold flows into my sleeping bag. I pull out my warm, pliable boots from the bag's bottom and then clomp over to the Craigheads' tent.

Soon Derek and I are climbing. Yesterday, while waiting on the ridge for others to arrive, clouds opened and I looked into the massive face of the fifth-highest mountain in the world. Its summit was 2 miles above, about the same vertical distance as to the Arun Valley below. In that view I realized the chance to see this mountain grow out of the night into the light, the sun flooding the summits of the world. As Derek and I climb, I calculate that Makalu's face has beheld 13,650,000 sunrises. Now he and I are about to share her 13,650,001st.

Though an hour still before sunrise, a glow ebbs behind over the planet. The two of us hurry as best as our pumping lungs allow. On Kangchenjunga, which is 20 miles behind us, a glow starts on the earth's third-highest mountain. As we arrive at the ridge crest, it feels like we are at a lookout atop the planet. Then light sparkles on Makalu's summit, like a bulb shining atop a Christmas tree, and starts washing the mountain in different hues: first amber, then yellow, draping the brown rock and white snow. Colours brighten until rich yellows flood the birch in front of me and warm my back.

In reality, the sun never comes 'up' in sunrise. The earth rolls bringing the sun into sight. As a boy after we'd returned to America in 1957, a neighbour several streets over, the unorthodox architect Buckminster (Bucky) Fuller, told me that the moment when the sun is seen from our turning earth can be accurately described as 'sunsight'. His and my friendship grew after I had once stumbled into his backyard. Thereafter, on Saturdays and afternoons

I helped him assemble an early geodesic dome. When the sun slips from view, he said, call it a 'sunclipse'.

Bucky was helping me see the earth in relation to things. One of his points was to understand that home is more an experience than a place--that helped me learn about my life turning and growing in me. Sunsight can be understood as an experience too: suppose we are riding a car through a dark street. As a corner approaches a glow may come from a side street, getting brighter as we approach a direct view of the light on that street. That is like the sun as the world rolls, and as on certain streets reflections can be seen off building surfaces like the false sunsights on those days when the atmosphere works as a mirror.

Today the sun preannounces her dawning. First comes the astronomical dawn: the world lightening in the east but dark still in the west when the sun is eighteen degrees below the horizon. Nautical dawn follows when horizons around and objects on the land are seen; the sun now twelve degrees below the horizon. Civil dawn comes when our world is lit but has no shadows: this is the time of dramatic reds against a dark sky, then warming yellows against a bluing sky. After that, here comes the sun, and with its warmth we say it's all right. And especially after a cold Himalayan night, its coming seems like it's been here for years.

Some time later, John, Bob, Dave, Tirtha, and Kazi arrive on the ridge. They pull out their cameras and take the morning pictures. Binoculars search back and forth, surveying the south-facing slope, stopping to look more carefully at exposed rocks, seeking the flick of an ear or any abnormal animated colour on rocks. The sun has been on that slope for three hours now and animals should start coming out.

'Hey look, a cave!' Derek points up-valley.

Why didn't I see this cave, I wonder, when I worked that part of the slope with my binocs?

'What a cave!' John studies it. 'Look, there's a stream not 20 yards from the entrance, a cave with running water. Grassy ledge in front for sunning. Damn, what a cave. There is a meadow in the valley below—all sorts of animals will feed there. Bamboo higher up for fresh shoots. Dan'l, how long would it take to get over?'

'From here, maybe three days. I'm surely tempted to switch towards getting there—the cave's an animal magnet. It must be the sanctum sanctorum for whatever animals there are in this valley. That hideaway is the place to look.'

'Wait, Dan'l,' John says. 'You're again sounding as though you're thinking about the chances of finding a Yeti here. You and I have been over all this before. A species cannot survive with just a mommy and a daddy. With the grizzly bear we've shown that it takes a population of at least two-dozen individuals. When numbers drop below the minimum for each species, it dies out. If you're interested in the Yeti, don't look in the jungle; rather, do the maths. You're not looking for one individual but for a population. And you need a habitat of adequate area to support a population.

'See the area of this valley. Two-to-three dozen bears are all this whole Barun Valley could support, and we're finding an impressive amount of bear evidence. If Yetis were here, a minimum viable population would also be about two dozen. And what evidence would such a population leave if it were here?

'That cave, though, looks like a bear's habitat. This valley we're looking into, higher than the jungle we've been in, is the first bear-like wilderness I've seen in Nepal. Why I came here—beyond the mystery of the tree bear—is to see how these bears are able to live with a habitat so reduced. Does the Barun give a lesson for how to pack a lot of bears into a small space? Around that cave is what I'd call a significant bear habitat.'

An hour later we descend to Makalu Jungli Hot'l. Traps must be checked, re-baited with smelly meat. A good stink, the Craigheads have found, is the way to snare a bear. However, as we descend, I wonder about that cave. Can a helicopter get there?

I will be using a chopper soon to start surveying this all into a national park.

꩜

BACK AT MAKALU JUNGLI HOT'L, SITTING BY THE FIRE, Lendoop and Myang turn the talk to the festival approaching in two months. Thousands of Hindu pilgrims arrive, especially Brahmins from the bazaars of Chainpur, Dhankuta, and Khandbari, for Barun waters are believed to be exceptionally pure.

'All are not pilgrims,' Lendoop explains to me. 'Myang is part of a group that sells *chang* or *rakshi*, corn cakes, and lamp oil. Others from our villages offer meals of rice and lentils. When Myang sells the jugs of chang his wife makes, or drinks it himself, he has fun.'

I quietly wonder how Lendoop with his debts will find money for the festival.

'Making money is harder now,' Myang inserts. 'Sellers travel festival to festival. The cloth sellers cheat us; they are not like the shopkeepers in Khandbari whose idea is to make you happy so that others from your village also visit their shop. These peddlers guess how many rupees each villager can pay. They keep the good-quality cloth in the front. But after you order, they give you the not-so-good cloth kept at the back.'

'The merchants work together,' Lendoop adds. 'Next to the cloth seller a tailor sets up; the tailors are low-caste people who carry hand-powered sewing machines and stich really fast. They never go over a seam twice; soon I have to repair each shirt or pants myself.'

'With the cloth merchants come trinket merchants,' adds Myang. 'Peddlers from India who sell beads, soap, cigarettes, candles, combs, and mirrors. When I was a boy from India, only pilgrims would come. Maybe this year I won't sell chang. My wife will, and I'll have a place for a free refill.' He and Lendoop laugh uproariously.

Lendoop continues, 'The good girls get drunk, and the best get presents. As the evening progresses, men join the chorus line, inviting themselves, other men join invited by the women. One singer leads with a song line, which is chorused by others. As the night deepens, couples sneak into the bushes.'

Pasang fills me in with the details. 'The girls walk past the trinket sellers very slowly as they head towards the dark. Sometimes their man then is helpful.'

Myang's turns to me in the firelight: 'You're not laughing, sahib? What's wrong.'

My face flushes as I reply, 'I was thinking about your wife. She is selling the chang.'

Silence slams into our once-happy fire circle.

But Myang quickly breaks the silence. 'Sahib, each of us has our own pleasures, which we may overindulge in. I have seen you take many things in excess. Maybe my wife is not happy selling the chang. You did not ask, but maybe she, too, has her own indulgences.'

'I'm sorry, Myang. I spoke too strongly.'

'No, I did not speak strongly enough.' Anger rises in Myang's voice. He is used to being obeyed not only by his family but also by his village folk. 'All of us overindulge in things we are fond of. You too, as I have said, ... having too much money. You ask whether I think about my wife. I do, but have you thought how we feel when we watch you have so much food, use so many clothes, and then pay us so little?'

'Yes, Myang ... and it makes me embarrassed.'

Pasang's teapot bubbles. After some time he ladles sweet tea into white enamel mugs, nicked and dented from a month of use in the jungle.

Each of us sucks from the hot cups. In drinking tea we share the communion rite of the Indian subcontinent. Tea is the common denominator shared by this land's hundreds of millions, the beverage that crosses caste boundaries. The fire flickers; its unifying flame bridging our diversity—flames holding us into its warm radius, while sipping from chipped chalices reunites us.

Pasang's son Tashi (a pharmacist, a painter, a superb interviewer, a promising ornithologist—mostly all self-taught) gently moves us into a less risky territory, 'Lendoop, some people say the caves here in the Barun are doorways to Shambhala, a valley of enlightenment. They say if you find one of those caves, you do not need pretty girls any more—the happiness of one night lasts forever.'

'Yes, I have heard of such caves. One is Khembalung in the Apsuwa Valley. It will take us a three-day walk from here.'

'Have you entered such a cave, Lendoop?' Tashi rejoins. 'Or known anyone? Did enlightenment come to anyone?'

'I have gone into many caves: holy caves and caves inhabited by animals. Not one of them was interesting. But maybe I did not go with the right frame of mind. Years ago, a lama from our village went into the Khembalung cave; he never came out. His metal bowl and a sleeping blanket were found at the entrance. Perhaps he entered Shambhala and enlightenment.'

I join, 'Lendoop, do you know any other times people entered such caves?'

Myang answers, 'I do. My brother had a friend who kept his yaks in the Mangrwa meadows. This is a wild valley with a big waterfall. One day my brother's friend fell asleep, and the yaks wandered away. Sharp screams awakened him. He looked to a cliff where he saw a fire burning in front of a cave and smoke rising from it. A shockpa and a man wrestled with each other by the fire. First the man would look like a shockpa, then, as they wrestled more, there was screaming and they changed with man becoming shockpa and shockpa then a man. However, suddenly the shockpa was thrown into the cave, and the man ran in after it. The cave door then closed suddenly, and the cliff ledge was now just a cliff.'

'Did your brother's friend investigate?'

'Although he was afraid, he went right away. But first he lit a big torch to protect him. Shockpas are afraid of light, you know.'

'A fire torch in the daylight?'

'Yes, shockpas are afraid of the power of light—that is why I carry this old battery,' said Myang, pulling the battery out of his pocket. 'That was why my brother's friend lit the torch as he climbed to the ledge.'

'What did he find?'

'Blood on the ledge; fresh blood in three small pools. The door to the cave could not be seen although he was standing where he had seen it earlier.'

My companions have all imperceptibly crowded in closer to our fire. 'Lendoop, do you know any caves up the Barun Valley?' I ask.

'Yes, but I do not know this cave in the Mangrwa. I have heard this story. I have talked to the Hatiya man whom Myang describes, but I do not know his cave. Myang has already told you, the door closed. The only big cave I know in this valley is the one below Makalu Base Camp. Near that cave a snow leopard in the rock guards the route.'

'I've heard of that snow leopard and cave,' I reply. 'Do you know any other caves up the valley? Maybe one with a stream nearby?'

'No.'

'Did you see that big cave this morning? Is there a trail that leads there?'

'I saw that cave. Never have I seen it before, and never have I heard of anyone going there. Somehow, today the door of that cave opened, but it was not there any time I've looked at that cliff. That is what Myang was saying. Doors open in rock walls. The mountains look steady but they are alive. Sometimes the mountains shake. You saw solid rock open today. Tomorrow if we go back, that door in the mountain may have closed. No, there is no trail to that cave. The mountains are alive with strange lives.'

Notes

1. *The Bible*, 'Book of Ezekiel', Chapter 34, Verse 18.
2. *The Bible*, 'Book of Jeremiah', Chapter 2, Verse 7.

thirteen

Bears and Bioresilience

December 4, 1984. Tirtha and I sit under a large maple. Near-at-hand, his plant presses wait for leaves now in two bulging knapsacks. Lendoop takes sprigs from the packs, then one-by-one unfolds the leaves on each, meticulously stacking specimens into little piles, holding down each pile with pebbles.

These are the last plants Tirtha will collect. To take its assigned place in Nepal's Royal Herbarium, each specimen must now be spread between newspaper pages and pressed in one of his presses. It is routine work the botanist has done thousands of times. His fingers sort, flatten, and order the leaves. There is a lot of time for us to sit under this maple, for finishing the leaves will run through tomorrow. Then our expedition walks out of this jungle.

Tirtha asks, 'Dan'l, you often mention your proposal, bioresilience. What brought you to this?'

'Tirtha, living with nature and being a mountain climber, I came to realize that one of biology's assertions did not fit what I was seeing.' Leaves flutter to the ground as Lendoop shakes a now-empty field pack and adds them according to type into the little stacks.

'After my doctoral work, I took some classes at the Yale School of Forestry. Class and books taught strength of ecosystems came from a diversity of species—and that diverse systems are more robust. But I'm a mountaineer. Climbers do a lot of thinking as they catch their breath and take in the views. And if you are paying attention to the biology you climb, it is obvious the higher one climbs that species grow tougher.

'I saw a contrast. Species numbers become fewer as altitude is gained, but each species is more robust. I termed this bioresilience. This name was in direct response to a word then used by biologists: biodiversity. Tropical biologists were arguing genetic diversity, many species filling many different niches, created biological strength. I do not argue that—but that is like measuring something by how wide it is without realizing things can be measured by how tall also.'

Tirtha smiles, 'Each looks at the world from his own perspective. Look what I do now. I am a biologist so I am defining the Barun by counting species. Collect samples, give each its Latin name; this my profession, and my discoveries show the biological richness of Nepal.'

'In that process,' I reply, 'you will show the lowlands of Nepal and the lower Barun have more species than up high. Your recommendation would then say: focus on preserving the jungles, don't loose those varied genetic resources. But there is a further dimension which is the hardiness of the alpine region where exist few species. Biodiversity and bioresilience, the strength of life itself is shown by both diversity of species and individual species strength. Two aspects of life, each complements the other.'

'And complementarity may be important,' Tirtha replies. 'For with biodiversity a whole species can go extinct with a change in temperature—and we now enter a world of temperature change. Niches tropical species live in are usually delicate. When that delicate world experiences temperature change, gaps can open quickly in that diversity supposedly so robust.'

'Precisely, having breadth in the system also brings less depth to each part. Shallow topsoil symbolizes this thin line of life. In tropical topsoil, which is actually very thin on the jungle floor, everything is alive, growing resplendently, but with that vibrancy is little dormant potential.'

Tirtha unfolds newspaper pages while organizing leaves. 'The complexity of the lowlands with their myriad species, plants, insects, birds. This is like a painting with a lot of colors spread out with nuance like a peacock's tail.'

'Precisely, Tirtha, but can a peacock live on a Himalayan peak? Tell me, you've been studying these slopes, is life more robust at the bottom or the top? Fewer species are higher, but those that are there can adapt to cold, heat, wet, darkness, able to respond to change, not just surviving but thriving. Consider just the tiny gentians; they are so delicate, but they take pummeling of cold, wind, and sun, and continue to bloom. They have a different luxuriantness. Changes that would extinguish life in the tropics (temperature, light, moisture) up high become dynamics of growth.

'There are two biological strengths, one richly fills the stable lowland and one robustly grows in zones of demanding life. The resilience of the hardy equips it to deal with climate unpredictability, adapting to cold and hot, building food reserves, enduring sun and long darkness—and when death comes, decay puts away reserves, allowing organic depth to build reserves.'

Tirtha laughs, 'As a scientist, whose training is understanding the tree of life, we look to classify species. When one is trained in beetles, that person is excited to find places rich in beetles. This is not questioning that value of biodiversity—only adding a second perspective, bioresilience, to describe another aspect of Nature's complexity.'

I reply, 'Since Linnaeus, the biologist's task has been what you are doing. Only recently have biologists started organized description of how species create ecosystems. Biodiversity one way to assess that focuses on differences between species, and bioresilience is another that focuses on abilities in one species to cross differences of habitat.'

Tirtha breaks in, 'Dimensions are more than biodiversity and bioresilience, for both are characteristics of species. Ecosystems, which you've just mentioned, you do not have quite right—ecosystems are how species engage with the resources on which they live. If resources are taken away, where does regeneration for the larger system come from? Ecosystems function in both biodiversity and bioresilience. In species diverse systems, regeneration comes from other life. In resilient systems regeneration comes from harbored resources. Habitat is definitive.'

Lendoop, as we talked though, was pondering another question. 'Sirs, why do you not kill animals; this jungle is full of good meat? We have a hunter of the King who carries one of the King's excellent rifles. No police will stop us from shooting. And yet, you eat old meat from Kathmandu. Why not fresh animals in the jungle?'

'Shikariji,' I reply, 'Certainly the deer here must be sweet. But hunting jungle animals will scare the bears. It is the bears we came to find.'

'Then why is Kazi allowed to shoot birds? Firing his gun five, eight, times each day scares more bears than the royal gun shot once or twice.'

Tirtha chuckles. 'You are right, shikari sahib. Maybe Kazi's noise is why we have not found bears.'

'I think you believe shooting is not good. Is it your religion? Is your caste too high? If so, let the hunter or me kill for you.'

Tirtha laughs now, 'It's not religion, Lendoop, but some of us are afraid too many wild animals are being killed and the world's wild places lost. The belief is like a religious belief. It is called conservation.'

'Tirtha sahib, I do not understand. The world has many jungles, many wild animals.'

'Not true, *shikari* sahib,' I insert. 'This Barun is the last wild valley in Nepal. A few more valleys are in China, India, and

Bhutan. And you may think the Barun will remain wild, but by the time you die, the Barun that you know as wild will have died *before you.*'

Lendoop looks at Tirtha. 'Doctor sahib, deer here are many. Bears are many, these damage people's crops, and little birds do nothing. Why do you who come, who possess so much, worry about animals that take care of themselves, that grow by themselves, when hard working people are without water, in villages with little food?'

Tirtha replies, 'Those who try to save the jungle do not live in it. Where they live, the jungle is gone so their focus is on that which is gone, animals, rivers. For aspects of the world that are no longer wild there is a special focus. And each of these departments has as its purpose a priority other than the full life of people.'

Lendoop interrupts, 'Our life organizes by caste. Is the Western way to organize by government departments?'

Tirtha and I both laugh. 'Yes, perhaps departments are the world's way to create caste,' Tirtha continues. 'In America few people are like the villagers of Shyakshila where the same person farms, builds his house, and goes into the jungle. People specialize as societies get wealthier. They call this specialization 'progress' as they work in their departments. That world of many niches they believe to be a stronger way of life. Mountain people are resilient, working across life zones, living by adapting, doing many tasks.'

I get the sense that Lendoop senses being spoken to as though he's ignorant. He rises and walks to Pasang's fire.

'Tirtha, perhaps Lendoop misunderstood, but I loved how you just characterized biodiversity using Western culture with multi-speciation into work niches and inability to adapt, then Shyakshila culture as bioresilient because it adapts across life niches. May I get you some tea?'

'Yes, thanks.'

By the fire, as I wait for the tea gives me a chance to joke with Lendoop about taking tealeaves to the botanist so he can study these.

As Tirtha and I drink, I then reflect to him, 'Conservation's work, Tirtha, as we were trying to explain to Lendoop, does not build to making a difference, despite intent to do. Rather, it helps our conscience.'

'What do you mean?' Tirtha asks, as with knee on a plant press he pulls the straps, adding two straps the other way then pulling these tight.

'Here's an example, Tirtha. Westerners see litter to be a problem. Litter pickup any group, schoolchildren, business executives, will mention as positive action. Clearing trash does not affect biodiversity, endangered species, or resource potential. Putting people to work thinking they are "saving the environment" masks the true problem which is consumption. Focusing on the little takes our view away from the big. Picking up trash lets people think they act on a problem when actually they are hiding it … hiding that they have trash in the first place.'

'You're missing the deeper lesson,' Tirtha replies. 'When we clean up after ourselves, we become aware of what we did. Do you not remember the awareness that came to you of your prior visit when you arrived here a month ago and found your old trash? Dan'l, your Yeti is not a search for a wild man but specifically for you, how wildness has left men.'

'How about the two of us take a walk in this still-wild jungle we will soon depart,' I smile. 'Let's stop pressing plants.' He and I refill our enamel mugs at Pasang's fire then step into the forest. Rounding a tree, we come upon Bob sitting alone, he gets up, and three men quietly walk on, together.

ೲ

WHAT BELONGS TO ALL HAS BEEN A CONCERN OF MY SISTER, Betsy Taylor, and her husband, Herbert Reid. While

13.2   Tirtha in Jungle

Source: David Ide

I focused on conserving specific places, they looked at dynamics planetwide, rethinking the ancient concept of the commons for the 21st century.[1] They introduce the point with a quote from Wendell Berry, 'the earth is what we all have in common ... it is what we are made of and what we live from ... we therefore cannot damage it without damaging those with whom we share it.'[2] The whole Earth is a commons.

Ownership, of course, begins in the mind: people believe some 'thing' belongs to them. Individuals transient on the Earth assume that existence outside their being is inside their control. Ownership of eternal continuation belongs to a temporal being. From that follows the right to control (even destroy) the eternal. The converse premise (the commons) is that we are connected to all. Global reality, Earth's very fabric, is being changed due to an idea that came into people's minds.

My sister and her husband argue that the commons is a concept from earlier ages that offers is a more authentic approach: stewardship in the present and across time. Stewardship differs from ownership; it does not posit that realities non-living, which have grown from Time's beginning, can be utilized by one part at one moment. The concept of the commons realizes that we come from shared inheritance, and in accepting the gift of today take on the responsibility to pass that on.

In assuming a fragment of the world that has existed for all time belongs not just uniquely but to a soon-to-die being who will consume that—the consequence, when ratcheted up, diminishes planetary inheritance. The loss created is not only that for today, but also across time, distancing us from not only each other but the lineage of time.

We move towards this declining potential, Betsy and Herb argue, through systems premised in the individual and that support the collective loss. We have created systems that move our global collective from an orientation of 'live within' to 'take from'. 'Products, ideas, images, money, and people zip more rapidly around the world. But, are we closer? Are we inhabiting and co-inhabiting more capacious, inclusive, and resilient worlds?'[3] The new view is a human-made world—lost is the perspective of common sharing.

In making this new creation, we step towards worshiping ourselves. In locating this within a system, that system evermore separates our awareness from the interlinked complex that is real reality. This gives me insight. Needed then, I realize, is to grow a new system not premised on individual ownership of the eternal, a system to grow the commons.

So to protect the Barun, our group must define the Barun in relation to people—a community that includes Lendoop, his king, our handful of experts, in an understanding extension to a global community. Our task is to create a commons for today and perpetuity, recovering a concept of sharing a world once all parkland.

Actions at one level will define boundaries, catalog species, understand relationships. This is the 'saving' step. This will require caretakers who abide by rules and make others abide. I do not yet know how to do this, but it is clear the answer must shun the customary option of management by paid wardens. Aside from increasing evidence around the world that this protection model succeeds only at limited scope, aside from parallel evidence that when effective it compromises the lives of those who lived near and were its stewards (like Lendoop and Shyakshila villagers), there is a pragmatic reason: neither I nor the His Majesty's Government have the money to add now in creating a national park for the Barun another traditionally managed park. I must find a new way to grow the commons, one for here and one that can be used in other places.

The premise of control must be let go. In the premise of the commons is the solution. Responsibility is with all. People will participate when they see a reason that is in their self-interest. Conservation that endures and expands is not pockets, is not protecting 'wilderness' (where people have been made absent). Success comes with setting up a system premised in living with, adopting and growing. For this is as with life itself. The search is not for a wild man but for how wildness has left men, then to bring that wildness back.

გად

WE SIT NOW AT THE CONFLUENCE OF THE BARUN AND ARUN RIVERS. Tomorrow the chopper arrives to lift us away. For two days we have been buying bear skulls from villagers and now possess fourteen. Each skull, except one, is old, brown, and dirty. Most have been sitting in villagers' granaries to scare off rats and mice.

'Well, John, what does our bear expert think?'

'Hell, I work with live bears, and they are grizzlies. Those bears reconsider their opinion about you in such a short time you can't consider them at all. In half a century of bear research I've never looked at fourteen dead skulls. Anyhow, let's arrange them into order. Some, you see, are tiny, others large; some old, others young.'

Over the next half hour, John arranges them according to size then spends more than an hour making notes about each, drawing up a little chart, with suppositions to presumed age, basing estimates on tooth wear and skull sutures. Then he lays out the fourteen according to the chart, makes notes as he sees relationships shift.

Our expedition group gathers around, 'Well, John?'

'I'm not a bear taxonomist. But, to me, all skulls appear to be from the same species. I see only size and age distinguishing. No feature suggests two species.'

'Maybe the fourteen are all tree bears, Dad,' Derek suggests.

'That is possible, Derek, but the villagers that brought them claim the big skulls at the end of the line are ground bears. At what point do we stop believing villagers when we have skulls, zoology's gold standard, in front of us? Further, what are the chances of fourteen tree bears and not one ground bear if two bears are in these jungles? Their trapping technique does not say 'yes' to one and 'no' to the other. With fourteen, probabilities are strong both are here.'

Derek comes back, 'Then, Dad, how do you explain their two bear explanation?'

'Keeping an open mind that maybe differences are there that don't show in the skulls, but arguing based on the skulls, I see only one explanation: tree bear is a juvenile ground bear. And, knowing something about bear behaviour, my explanation to what villagers experience is that sonofabitch old male bears, seeking to dominate food, drive small juvenile bears up into the trees where these big males cannot follow. Adult aggression makes the big bears "ground bears" and the little bears "tree bears". When a mother's around on the ground, the young are protected, but once a cub leaves mom … when papa reappears, the yearling or twoling skedaddles up a tree.'

We absorb John's proposal of old-man *bhui balu* and twoling *rukh balu*. It seems to make all the reported differences come together. Aggressive bears are adults and big, the shy are young and small. Villagers' behavioural claims are explained, as well as the absence of distinctiveness in skulls.

A thought comes to me: 'John, in some animals, the butterfly for example, there are major differences between juveniles and adults; a caterpillar is totally different from butterfly. Among bears, do other juvenile and adults differ in behaviour as you're suggesting for tree bear and ground bear? With such, might you be suggesting something unique to these bears? Between the two, we're hearing claims of big differences in food type, habitats, as well as behaviour. Among humans, of course, such differences exist adults to children. Does such exist in any bear?'

'All bears I know behave and eat essentially the same juveniles to adults. Maybe you've popped my hypothesis. It'll be quite a discovery if in a nonhuman species is such a difference—but remember humans developed our age-based differences recently: parents go to work and children go to school, boys wear pants girls wear dresses, even wearing clothes is worth noting, a trait, found not in our genes but in the phenomenon we call civilization.'

Tempted to suggest we take off our clothes, I reply: 'John, I don't think you should be surprised about this bear being like people. After all, it has a footprint like a human, we even saw it using snowshoes it made out of rhododendron branches; people have called it a snowman for a hundred years.'

John laughs, 'Hell, you never let up! Let's break camp and get to hot showers.'

'No John,' I grin back, 'let's return to these valleys and find that cave.'

13.3   John in the Jungle

Source: David Ide

༄

I AM EN ROUTE FROM THE USA TO THE HIMALAYA
AGAIN. I've been making this flight when in my first trip
in 1947 *en route* meant not airports but also water ports and a
'flying boat' with splash landings in the Nile, Arabian Sea, and
Karachi Harbour. Intercontinental flights require turning activity
to the mind, and the crutch to make time pass are books to let the
mind go to other places. My early trips were having read to me
Peter Rabbit and *Make Way for Ducklings*. Later came my reading
gritty cowboy novels, then *National Parks, and Protected Areas:
Priorities for Action* by the same Jeff McNeely who earlier chased
Yetis. Now I fly with Bill McKibben's *Eaarth*, and Jack Turner's *The
Abstract Wild*.

Conserving the Barun will require bringing *Homo sapiens* in, in
partnership. We who gather to do this have little money and are
not a big organization—yet with a people-based vision the objective is not protecting a valley only but connecting to the world.
People are part of Nature. This is equally true for pristine jungle
as the remade nature of cities where people, even in walking never
touch things other than human-made. Jack Turner is telling me
the wild is still possible, in the Barun and in the manufactured

world. With people as partners, scope changes because no longer is the action removing people.

This story of people's actions with the wild is being written in the sky also, for flowing out behind my airplane extend contrails. They are a line across the sky for a short while. Then they melt into air like footprints in snow. But extending from them are imprints not seen, arcs of carbon. Here, higher than Everest and Makalu, are signs in substance that began within the Earth not long ago, made from earlier living much much longer ago, where their contrary tales will last long into Time. We have made a new Eaarth, McKibben is telling me.

Communications are what binds our new life ways. The new Eaarth is fundamentally one of hyper connection, an info-sphere being melded with the changing biosphere driven by an econo-sphere. Binding linkages have drawn these spheres together. Over receding horizon behind extend radio waves, and magnetic waves pull us to the horizon ahead. The new info-sphere knits together the bio- and econo-spheres—knitted in a socio-econo-info-biosphere.[4] Changing now is presentation of wildness.

The old spheres in these realms were separated. As a boy I thought what defined the wild were animals that could jump out and eat me. But wild animals are not the wild; they are individuals like myself who present as life I cannot control. Potential now for living in wildness is everywhere. We must open to a world we do not presume to control. It allows two people to talk without speaking while walking a night trail; it speaks to the people of Shyakshila of my coming, and it reaches to deep recesses of humanity that have never met and causes both Jennifer and me to identify kin in a girl in a basket *and told the same to her?*

I need to learn to navigate because I am part-of, one-with, using-all-of the world. Connection will carry me, not to things of my making, but to life that I can thrive in but do not control. Doing so I can mimic the more skilled navigation of birds that fly miles above, whales that navigate leagues below, or elephants that

13.4   Kazi Using Mist Nets to Collect Birds

Source: David Ide

remember across generations. I can join the greater wild of life that does not seek to control. Each in its place is the constant surprise, journeys into the new wild of the new age.

My growing understanding is that dimensions go beyond space and time—as even an airplane flight lasts in its residue in the air *and also* in its repositioning of people, long after the airplane lands. The living lived continues on for years after death—not only in our children and children's children, but in what they made of and with their lives, ripples of being in the space of the evolving socio-econo-info-biosphere with consequences that affect the whole and reshape the world.

෬

THE REALITY OF THE EMBODIMENT OF IDEAS. As unlikely as Yetis for which exist footprints, more relentlessly embodied across the ages is the proposition for angels. Belief in angels endures, not as carriers of the wild but as carriers of Divine. There are obvious differences, but also exist similarities. Are angels actual beings who flutter to our world, or idea shards from ancient belief? Either way, they still speak connecting to the world outside human control. But outside the airplane window I see dawn breaking.

Understanding comes to our souls. After the understanding comes, subsequent facts confirm. Was the understanding truly from another being? The sceptic may scoff that such is possible— for we have not seen it, measured it. But a message came, and subsequent facts confirmed. So like footprints it is there. Like the Yeti that speaks of people from the wild, so do angels from whatever this existence is beyond human control and knowing. Messengers come from that which we cannot see and measure. Dismissal is harsh, for ask that sceptic about the reality carried with falling in love, of giving that empties oneself for another. Or, ask that person about their mother. An angel can indeed come, though it cannot be objectively confirmed.

The imprints of this are so real they can (and have) turned human actions. Across generations, in the transformation that follows is the reality; reality is not in the messenger but in the result. Some 'thing' was conveyed. Because earlier generations described the messenger as with body does not require we accept that part of the understanding. We can accept just the understanding that was conveyed, through eyes or ears for some, for others through witness.

What resulted was moving whole cultures towards grace or disgrace. Deep knowings told how to walk life trails. The results have endured. They carried aborigine songlines across centuries, allowed people to live in conditions where Modern Man could not. Conveyed by them are knowings that opened understanding of the impossible in a world where senses do not transfer. It is not Science, but it is real. It is real because, from such knowing, results can also be repeated and objectively verified.

Names are offered—Jehovah, Allah, Gaia, Great Medicine—expressions of a being whole and alive. What matters in such greatness is the enduring message rather than the name. A transcendent was made alive, gathering the whole into changed parts. Such embodiment is also shown in geniuses (even the hyper rational accept this ability). A message comes out of the not-previously-thought-of. An arabesque of association linked domains previously unconnected. A message transferred.

This meta phenomenon now dabbles with other explanations also, seeking deep understanding of the 'uni' in universe connecting to the bond that makes our self ourself. While the force that does this may confuse, the result is clear. The bond creates group, and from that individual identities transform among people. (This also happen with whales, or ants, even blades of grass.) Living directs to living with the whole. We accept this among people, also in a pod of whales, or even ants in their trails, if not yet willing to accept in adjacent grasses growing.

This search has been explored across many ages. Recently helping understand those lines, we have String Theory. Vibrating strings of energy, it is alleged, reside at the core of all matter, flexing and reshaping, the most basic of all embodiments, a matrix linking material definition. Their peculiar vibrations create all that matters, presenting too in dark energy behind the reality we know. Might those vibrations connect to other universes—creating not one universe we live in but more accurately a multiverse? In the distant cosmos energy pinpoints speak in such verse. Instruments can measure those points. Does the soul go inward to another nested set not measured? Might our inner senses also be monitoring to that through our strings to Life?

Ancient songlines are in such searches. From them arise portraits of the present, identities more vibrant than portraits shared in electronic digits. After we no longer relegated to religion alone connection of soul to the great beyond, authenticity of the scientist's voice ascended. Ordering allowed verification one investigator by another, going beyond postulate to proof. Greeks as well as Chinese were early explorers of the scientific method, but a good year to begin systematic life ordering is 1753; the name is Linnaeus. From a Babel of folk names, all Life was positioned: kingdom, phylum, order, family, genus, species. From an alignment where strings of DNA reported to others in presumed sequence, then followed a search that sailed the world to order of Life.

Name and position gave partial understanding. The extent of its partiality shown in an insect colony in a rotting tree: life through death becoming life. Brain cannot explain how animals without reasoning can create such. But complexity theory does. Explanation comes in parallel understandings through the recursive mirrors of folk knowledge, poetry, art, and religion—life looking into itself and seeing back the wonder of it all. Seen is each mite, giant panda, tropical fig tree, dependent on another—understood not as a tree of life, factoids, but as wholes, more than part summations when all together.

Eighteenth century biology was an organizing of species. Nineteenth century biology gave new understanding when Darwin and Wallace formulated natural selection as the speciation mechanism. By the mid-twentieth century relationships were recognized; *ecology* was nets holding Life, and lexicon added a new hierarchy: organs, organisms, families, communities, ecosystems, and landscapes. No species was alone, no system independent, mite, panda, or fig. Two taxonomies intersect: genetics and of relationships. And despite embracing parts as well as relationships between the parts, neither defined the ability to respond to change.

For our response can unfold in so many ways, bringing me back to bioresilience: predation, nutrient changes (especially food and water) and habitat fluctuation (temperature, precipitation, and habitat loss). Response comes through defence—protection (adding a barrier or insulation, or creating a storage mechanism); adjustment/mutation—changing to accommodate the change (reducing consumption, changing metabolism, tool use, or altering reproduction); or migration—moving from the change (temporarily or permanently). The challenge I must solve is how to move among all these.

As the warming planet unleashes unreliability, zones that have not experienced freezing for 12,000 years will have vascular systems ruptured, their wonderfully intricate life-carrying pipes broken. Other places will be scorched. A wild is coming augmented by human action now beyond its creator's control. In this new world, consider the magnitude of humanity's extension. If the mass has their weight combined of all mammals that people possess (cattle, horses, camels, dogs, pigs, chickens), this would constitute 65 per cent of all animal life. Weighing all people will constitute 30 per cent. The remainder, the weight of all wild mammals, will comprise 5 per cent. The extensions of humans and their animal dominions comprise 95 per cent of mammalian life.[5] One animal has nearly annihilated the wild. Vaclav Smil who made these calculations suggests we approach an ecology designed and managed by human intelligence, a noosphere.

As we try to frame this noosphere, we need to expand the concepts. Bioresilience helps, a way to understand and measure accommodation to change. It allows, for example, us to understand biology's ability one moment to promote rapid photosynthesis as sugars and moisture surge in vascular systems, then reposition almost instantly as sun departs and freezing blasts in. Or, it allows to understand the needed ability of life to harbour nutrients in roots and rhizomes so to bridge food deficiencies when rains do not come as we risk annihilation of life.

For in addition to measuring the diversity of genetics, we need to track species' abilities to reposition. Mountains are one such waiting place where biome stacks on biome, networks poised to move up, down, or outward as change dictates. Another set of places so waiting with bioresilience is the typical town park, places in our midst that endure collective trauma of pollution, precipitation, and human populations, dynamics that devastate the pristine.

Town park species may not be as beautiful—geraniums are not as compelling as orchids and many warbler species are needed to encompass the food range of the crow—but town parks are seed beds of resilience as national parks struggle to preserve the exotic.

This is the Anthropocene where we redefine the wild as well as the world. In fundamental ways we approach reconfiguring the modalities of evolution. In it, the most resilient of species will—whether intentionally chosen, or, in absence of choice, evolve the result—build biological bridges. We shape the processes by which we go forward, men, seeing to avoid the abominable, but creating the new wild. When the massive extinctions came at the end of the dinosaur era, another species having great appetite, had no bridges to carry them to the new changed world. But humans have the ability.

Notes

1. Herbert Reid and Betsy Taylor, *Recovering the Commons: Democracy, Place, and Global Justice* (Urbana and Chicago: University of Illinois Press, 2010).
2. Reid and Taylor, *Recovering the Commons*, p. 51.
3. Reid and Taylor, *Recovering the Commons*, pp. 8, 9.
4. Daniel C. Taylor and Carl E. Taylor, *Just and Lasting Change: When Communities Own Their Futures* (Baltimore: Johns Hopkins University Press, 2016), pp. 6–9, 32–3.
5. Vaclav Smil, *The Earth's Biosphere: Evolution, Dynamics, and Change* (Cambridge, MA: MIT Press, 2003).

fourteen

Entrapping the Yeti

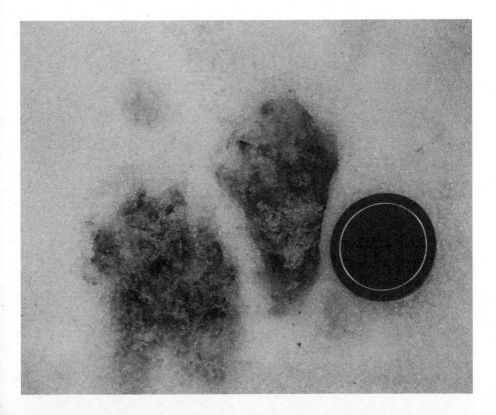

Previous page:

14.1   Side-by-Side Positioning of the Hind Foot (Left) and the Front Foot (Right) with the Foot Size of *Ursus thibetanus* Shown by the 52 mm Black Lens Cap

Source: Author

25 October 1985. Ten months ago, leaders in Nepal were excited that their country might be introducing a new animal to the world. What defines a 'new animal'? An animal new to science is the usual assumption. But the excitement a year ago was pride that Nepal's people had made the identification; it was a validation of local knowledge. What was found, then, was more important than an animal—it was the Barun Valley itself—that the country had an unknown treasure. The discovery had come (as described in Chapter Twelve) as Tirtha and I had sat that afternoon on the Barun ridge looking out, and he had realized then described what his country still had: 'The Barun shows the wet Himalaya before people changed the land. Indeed, it shows the wet jungle of Asia ... A view of what would be 5,000 miles on the flat can be traveled with a turn of the head ... showing Asia's habitats between tropics and arctic before people came to this land.'

In this identification by Nepal's leading botanist what was discovered was a heritage, a legacy larger than one species. A breadth of wildness was revealed—and it was Nepalis (not just Tirtha, but

Kazi, and others, not the least of which was His Majesty who had known the Barun was 'the most wild'); it was a community of Nepalis who had discovered this. For years what had excited me was a Yeti explanation; to answer that the Craigheads had joined my search. But the big discovery our team had made was Nepal's still intact wildness. So now I walk with my friends through the hallways of the government to introduce a new way of thinking about protection based in people, conservation through the engagement of people. Two weeks from now, international participants will arrive for a symposium worthy of their busy schedules, to outline the structure of a new national park that will continue this partnership, an initiative that grew from questions about a bear, and now our idea is to ground this park in management with local communities.

Tirtha, Bishou, and I sit over afternoon tea to further evolve that idea. 'I promised the international visitors a chance to see the centre of the world's highest mountains. Can we move our planning meetings to include Nepali leaders with the international experts? Can we take the ministers and officials to go sit with local leaders and do the planning in bamboo huts?'

'We've never had a conference like that—one in the heart of the Himalaya in partnership with the people!' Tirtha is enthusiastic. 'Our leaders are mostly bored in offices. Everything depends on how the invitation is issued'.

Bishou lights up, 'Tonight is the Queen's birthday party. Before that, let's invite His Majesty's uncle General Sushil, Chancellor Bangdel of the Royal Nepal Academy, and Vice-Chairman Sainju of the National Planning Commission. If news is out this group is going, at the birthday party tonight others will want to join. We then control who comes using the limited seats on the helicopters, and in controlling who comes we shape the agenda to trusting local people'.

The following morning we gather again at Bishou's house. As he serves spiced quail eggs, Bishou is excited. After the party he met with Prince Gyanendra, 'H.R.H. called me to his palace before leaving for Thailand. He knows World Wildlife Fund has trouble because it did not involve villagers in the Annapurna project. He wants to push a people-based approach. At his command, already this morning I telexed the Chief District Officer in Khandbari to tell him we were going to have a people-based meeting. I also called the Royal Flight Wing to reserve a third sixteen-passenger helicopter'.

Tirtha is worried. 'Last night some people wanted to know from which districts land will be taken for this park. Others wanted details of people-based management. A few hinted that involving people might weaken the monarchy'.

Bishou breaks in, 'Keep ideas simple: A new park. The objective is a people's park in Nepal's finest jungle—don't say how'.

Tirtha is still worried, 'The international voice may be a problem. We do not want another example of foreigners telling us what to do. The people's voice needs to be authentic'.

Bishou is not listening, 'Okay, we agree about the park. We must not let this meeting raise questions. Present an idea that has no problems'.

Tirtha turns to me, 'Dan'l, how much money do you have? Right now, we're talking forty people in the jungle for several days. Helicopters will be expensive. After that there will be big money'.

'I have USD 15,000', I answer meekly. 'Last night I figured fieldwork for just the boundary survey will cost USD 25,000. Now add in all these helicopters, and a lot of time in the field setting up meetings'.

'You have months to raise the boundary survey money', Tirtha answers.

'I have the USD 15,000', I answer meekly. 'There was a nice man in Aspen Colorado'.

'You need a lot more', Bishou admonishes. 'His Majesty said you could be trusted. You must stand behind what I have promised'.

Gently Tirtha pushes, 'Dan'l, you're asking my government to make a perpetual commitment. Then you started talking about changing how this government views ownership. When you talk such ideas with important people, you need to do your part'.

I lower my teacup. An image flashes of two sceptical trustees at the Mountain Institute's next board meeting. They're going to have questions and want money locked-in. But I cannot stop this building-while-doing. If I did, I would have to leave this country today—and never return. As the organization's president, I can call on a credit line at the bank; that will pay the chopper and jungle bill. After our meeting, I will have a plan as well as government support. On these I can raise money. I look at Tirtha, 'I was telling you about money I brought—I'll wire for more'.

Bishou jumps in, 'I told His Royal Highness last night that his brother said you can be trusted. Here is the new problem. We must have meetings in Kathmandu to get people involved who we cannot be taken into the jungle. I decided we will rent a hotel, get the news media. What do you think, Tirthaji, will two days satisfy everyone? We must serve food and strong drinks—tea and biscuits will not make the project look serious'.

Tirtha nods. 'Two evenings should be enough. People do not like all-day meetings. When including people they'll want to meet dignitaries. No need to announce that the King is behind it; they all know that. We must be open so people feel that each can join in'.

My head spins: egos of Kathmandu intelligentsia … tea and biscuits not enough … a hotel ballroom for two nights … newspaper, radio, television. A minute ago, I thought I was approving helicopters and jungle staff. I'm seeing now the difference between chutzpah and maybe going out of business.

Bishou follows, 'We need another meeting afterwards. First two nights to launch the idea. Then another meeting at the end with the public endorsing the jungle meeting. Everyone will feel they have a part but understand that there's not enough room on the helicopters. Yes, another meeting when the jungle group returns'.

14.2 The Delegation Arriving by Helicopter for the Makalu–Barun National Park Planning Meetings at the Saldima Meadows

Source: Author

My mental cash register keeps ringing up costs: helicopters for twenty Nepalis, nine foreigners, tents, camp staff, a third grander reception in Kathmandu. I could look for the Yeti with a backpack. But starting a national park demands a strong stomach.

Bishou again jumps in. 'That final night should be at the Royal Cottage in the King's Forest, what do you think? It's only twenty minutes from the city. People will find it interesting and that will bring more to the meeting. The cottage is under my control'.

My cash register keeps guessing. This approach is what I believe in—the energy generated will be perfect. I'm going to have to be criss-crossing America raising the funds. But to do so I have something better than a skull. I will have a national park in the world's highest mountains, not footprints disappearing in the snow.

RIDING HIS MAJESTY'S HELICOPTER, ROTOR BLADES TWACKING, this flight was arranged by Bishou as a training flight before carrying royalty the next day, so this day's bill is just for the fuel. As we fly through the valleys, I look at the houses hung onto the hills, remembering how Wendell Berry summed up such tasks: 'The forever unfinished work of our species ... the only thing we have to preserve nature with, is culture: the only thing we have to preserve wildness with is domesticity'.[1] Nature is not going to be set apart by what we are headed to do, rather people shall be rethinking how they live.

The helicopter is over Makalu Jungli Hot'l and, remembering the weather, I am chastened that my biggest risk may not be financial. Once we land, we might be cut off by the snow, and problems will escalate if I cannot get these people out. This conference could well become a front-page-news disaster. In organizing this meeting months ago, knowing that risk, I selected the week in the year I believed least likely to bring stormy weather. But it is impossible to predict the Everest/Barun weather.

Swinging through a steep turn, the helicopter lands. Pasang, Tashi, and Lendoop run to us. Two days before, after a two-week walk, they brought in a string of porters with food, field camp supplies, eight chickens, and dragging two goats over mountain passes. Tents are up, and they've been at work building a conference centre. The yak herder who arrives in a month will find his former hut with sweeping woven bamboo wings. One wing, twice as long as the other, has workers weaving bamboo into a conference table. A glance to the stream shows stones creating a pleasantly lined, very public washing area. Men will accept that, but will the women?

The next day a helicopter arrives, then a second. On landing, everyone's head turns to the waterfall spurting out of the base of the fifth highest mountain in the world, the cliff Lendoop once described as: 'Imagine a great mountain urinating'.

John Craighead spots something beyond. 'Dan'l behind, is that smoke? Who could be building a fire that high?'

A thin wisp rises beyond the waterfall. John brings over the NASA satellite image he brought, indicating a large frozen depression behind the rock cliff. Around it are snowfields; these must feed the waterfall. The smoke seems to be coming from inside that depression ... if it is smoke.

Twenty-one people, twelve senior Nepalis and nine Westerners, are in the Saldima Meadow; with them are six local government representatives. All of them are used to being busy with meetings augmenting tightly scheduled workdays. For people like that, a meeting such as this requires around-the-clock activities, morning bird trips, evening campfires, and snacks.

General Shushil Shumshere Jung Bahadur Rana, His Majesty's uncle, opens the seminar. 'As chairman of today's session, it is my pleasure to welcome all. How unusual to have so many distinguished people....' Just then, Bill Garrett, Editor of the *National Geographic Magazine*, snaps a photograph of Scott McVay, director of the Dodge Foundation, squeezed on a hand-hewn bench between Nepal's Secretary of Forests and the Chancellor of the Royal Nepal Academy. The flash slips immediately out of the cracks in the woven bamboo. The meeting continues, our chamber lit by pinstripes of sun slipping in.

'So many distinguished people in such a humble setting, the conference table was lashed by villagers who cannot read, who have never seen wheels move a vehicle. Their skills differ from ours. But if they know the branches with which to build a table around which to talk, we hope they will bring knowledge about how to care for this jungle'. He turns to the local people respectfully sitting apart from the table, 'Please join us at the table, and I ask you also to speak. As we talk ideals as well as ideas, let us remember we have come to design a park based on people. What more of an appropriate setting and how much more qualified a group. Ladies and gentlemen, welcome!'

In deliberations that followed, a member of the royal family carried food. The leader of the National Planning Commission went around camp picking up trash. Whatever the bear's identity, the discussion had moved to the people. The politicians outline what might get through the government. The scientists tell what must not be forgotten. The bureaucrats wonder how to process the unfamiliar. On flip charts, a core zone of 300 square miles is designated connecting to the territory of Everest already protected. Circling it will be a conservation area open for non-destructive human use. An adjacent World Bank hydroelectric project is noted that must be considered. Rapid population growth of the surrounding villages is also noted, developing plans for how to deal with these communities who, as populations and ambitions grow, expect to move onto 'now unoccupied' land and displace the forests and animals we seek to protect.

Like all places in earth's Anthropocene, this place is not 'untouched' by humans, and our discussion intends to steer human touching that will soon become more. As the planet adjusts to its most destructive animal, our ambition seeks to allow this place to be adjacent to human impact, a little more with natural influences.

Tirtha uses the phrase, 'We must strengthen Nature's beguiling natural mask, to make this place touchable'.

On the last night Bishou announces we're going to have a bonfire. Three bottles of scotch have somehow appeared. To prepare for the party, three old friends leave the Saldima Meadow to fetch firewood from the jungle. John and Tirtha sit on a fallen tree in Nepal named after a man from Scotland, *Rhododendron campbelli*. I sit across on a rock, 'Tirtha and John, are we starting an approach that can transfer?'

'That we must see', John says. 'A park where Man is managed as one species, as part of the ecosystem sounds good, certainly better than a hundred years ago'. John is referring to how the US Cavalry rode into Yellowstone, threw out the native peoples, then with a military focus on borders secured the perimeter to create the Yellowstone National Park.

Tirtha follows, 'We have involved the right people'.

'And what must be kept is that sensation which feels bigger than the area itself', John says. 'Ecosystems are more than the land; they compress a wholeness into one place; they make you feel small'.

'John, you're sounding mighty Buddhist', I chuckle. 'What do you think Tirtha? Doesn't that sound like finding the ocean in a single drop, except that now it is the earth through a single park?'

'It is more complicated', John says. 'A grizzly bear is more than the animal. Locking it up in a cage sets it apart. The Buddhist explanation about that is reality in existence as reflection of others. The relationship between a caged and a wild animal is of wilderness and civilized. A grizzly in his home range outside the cage presents wildness bigger than the animal itself'.

'Okay', I break in as I want to change the conversation. 'Here is a question. I recall one way to tell whether a meadow is pristine is to examine the proportion of grasses to sedges. Yes?'

'Why ask that after just designing a national park?' John asks, mystified.

'John, usually when Dan'l asks strange questions he is thinking again of his Yeti', Tirtha smiles.

'Tirtha is right', I say. 'But please tell me about grasses and sedges'.

The botanist begins. 'A pristine meadow usually has few sedges. Grasses are taller. So, if the meadow is not being grazed, the grasses block the sunlight the sedges need. In a truly pristine meadow you may find only fifteen per cent sedges. But when the meadow is grazed, especially with domestic animals, grasses are cut back, the hardier sedges come forward, and you may find only fifteen per cent grasses—as well as more wildflowers'.

John adds, 'Fluctuating habitat seems to bring the hardier sedges'.

'Sedges and grass, how do we tell them apart?' I ask. 'Both look like grass'.

14.3   Meeting in a Bamboo Hut to Plan the Makalu–Barun National Park: (from left to right) Bill Garrett, Editor, *National Geographic*; Beverly Osman; Scott and Hella McVay, Dodge Foundation; and John Craighead

Source: Author

'True grass has round stems', John replies. 'Sedges have triangular stems and flower from one point whereas grass flowers from nodes along the stem'.

Tirtha continues, 'The grass/sedge issue is interesting given the bioresilience idea you struggle with, Dan'l. Many people think pristine systems hold the maximum diversity of species. But pristine systems can climax out, and a dominant species takes over. Biodiversity is maximized usually where habitats are in flux, for example, forest and meadow, or the estuary interface between land and sea, as well as when people are responsibly using the land. Multiple niches exist at such interfaces, and multiple niches propagate biodiversity'.

'I'm still working through how grasses and sedges could, in any way, relate to your Yeti?' John asks.

'Let's pick up these dead limbs and return to camp', I reply. Tirtha laughs and we drag branches up the cobbled trail, a trail used occasionally by smugglers taking the back way into Tibet.

∾

OUR LAST MORNING IN THE CAMP. DAYLIGHT HAS BEEN WITH US for an hour and there has been more activity than on any morning before. Clothes are being jammed into duffels, sleeping bags stuffed, breakfast hurried. Scott and Hella McVay slip off for a last walk. I lean against a rock, writing notes and look again at the Saldima waterfall thundering out of the rock face. Today, wisps also rise from behind us. Is it smoke?

But uncertainty hangs over the camp. When will the helicopters arrive—if at all? Worry is not here in Saldima for there is blue sky above. In Kathmandu, November mornings are heavy with fog, sometimes not lifting until nine. The flight here takes an hour and ten minutes. The bottles of scotch last night got everyone talking.

Bob Davis comes over. 'Dan'l, the pattern is holding: look down valley. As we said last night, this valley is clear until about nine-thirty. Then that cloud down there starts rising'. Bob points to the lower end of the valley. 'The cloud now is at 6,000 feet. Those choppers not only must get in by nine-thirty, but since we need a shuttle to position people outside, they must get in twice. So the fog in Kathmandu must lift by eight'.

'Hot tea, sahib?' Pasang holds two steaming mugs, then turns to the hut he's made his camp kitchen. Fatigue in his voice tells he's ready to leave. Cooking for so many dignitaries, and each with likes and dislikes, having few resources, was a strain that everyone has ignored but he's managed to carry. People like Pasang get noticed only when something goes wrong.

As he steps away I ask, 'Pasang, how many more days of food do you have?'

'After today's breakfast, three more meals, but sugar for only two more teas'.

Twenty-one dignitaries, six local officials, three kitchen staff, and twelve porters kept as backup in case of trouble—in all thirty-seven people. When planning, we brought what we calculated would last eight days (five days of meetings plus three of reserves; it turns out we have a day and a half of reserves). I sip my tea, Pasang as always made mine extra sweet. Others must have also gone and asked. Sixty pound of sugar were used.

Bishou comes over, 'We are successful. Our idea has the soldiers now'.

'Bishou, we don't know if friends of today will stick when they return to their offices'.

'Yes, we do not know who we can trust', he replies. 'But I think we have enough. Look. Have you ever seen so many powerful Nepalis together for such a task?'

Dan Vollum, a conservationist from the Pacific Northwest who is also a helicopter pilot, walks up. 'That chopper better slide in soon. Flying up the valley among those clouds will be like navigating up the belly of a snake—curving walls ready to smash the bird at every turn'.

Bishou smiles. 'Mr Vollum, don't worry, General Shushil is here. He is in charge of the Royal Flight Wing. The pilots will come even if it is to fly up the belly of your snake'.

The cloud is almost on us. Maybe the chopper can hop over it. But just then, from inside the cloud comes an unmistakable thwack-thwack-thwack.

Dan Vollum points, 'See, there!' Following his finger, I see a giant bird, made in France, flying in the Himalaya at ground level, not in the cloud but below it, not flying fast but bush-by-bush; disciplined pilots coming for their general.

That chopper will soon exit out the blue hole above. One group will get out. Then we see a second chopper wallowing in behind. People surge towards them. Our rations just increased by two-thirds. The Nepalis step aside for General Sushil. I rush up, waving my arm to speed the departure. 'General, please. It is time for you to leave.'

The pilot looks expectantly. Brigadier Sushil motions the others on board and pulling me away says, 'Dan'l, I stay. You are low on sugar, I think'.

ೂ

MAY 1990. MANY OF US MEET ACROSS THE BORDER IN CHINA near the northern bases of Makalu and Mount Everest. We're in Shegar, Tibet, in the People's Republic of China to launch a nature preserve that adjoins Makalu–Barun National Park in Nepal. This Chinese preserve is ten times larger than Makalu–Barun, three times bigger than Yellowstone; or, as our Chinese colleagues point out, larger than the island of Taiwan. This is the largest preserve yet conceived of in Asia. China likes big.

Across its 180-mile width, it adjoins five parks in Nepal, then with a brief non-protected gap, a park in India's Sikkim, a series of national parks in Bhutan, and on to India's Arunachal Pradesh, then Myanmar, connecting protected areas across a 500-mile expanse of the eastern Himalaya. One-third of the breadth of the Himalaya is being protected, a swath roughly 500 by 20 miles. In time, preserves in southern Tibet, collectively known as Four Great Rivers, will continue to link to preserves in Yunnan and down into Cambodia. Following the Saldima Meadow meetings, discussions spontaneously began in each country. Nepal changed its national park approach. China adopted a new concept. India and Bhutan started conservation rethinking—what drove this parallel discussion was prioritizing conservation for the local people's benefit, not nature's sake.

So actions began to improve both protection of the wild and people's living. The easy-to-understand example of wild meadows growing more flowers when domestic animals graze parallels how nature can become more beautiful as people use it. The result is conservation unfolding at the landscape level. Protected nature is not only keeping places untouched (it is), but also a growing understanding of how to touch the world and help it grow.

Land in this now-evolving understanding is set into zones allowing protection and use, balancing zone-to-zone across a region that is now termed 'landscape' within which there are varying nature-sensitive management policies. One zone is the 'core area' with not-to-destroy guidelines, a second 'buffer' zone allows actions such as animal grazing but no human inhabitation, a third zone is 'agricultural' with fields yet also parameters of nature protection, a fourth zone is for denser human settlement (towns, even cities) where destructive impact to a pristine wild is confined and managed. From what the Government of China has called Qomolangma (Mount Everest) National Nature Preserve (QNNP), areas in Nepal, India, Bhutan, Myanmar, and Cambodia extend a concept for all of humanity.

Following the meetings in Lhasa we have just initiated, the Government of China has decided to make the QNNP a 'national treasure' on the same protection status as the Great Wall and the Ming Tombs. Underway now is a growing international partnership. Here in the centre of where 40 per cent of the world's people live and where the highest points of the planet stand now advances arguably the highest need on the planet: earth's responsible care.

To push this forward, the partnership on this trip has come from four countries. While some in the larger world press for Tibet to be free from China's rule, our actions work beyond today's politics to build a platform to grow solutions for humanity's longer problems: human life requires that the environment be protected. On that base will grow economic growth, health, education, and governance. A participatory approach in conservation is our objective to grow bottom-up capabilities among people for later larger governance.

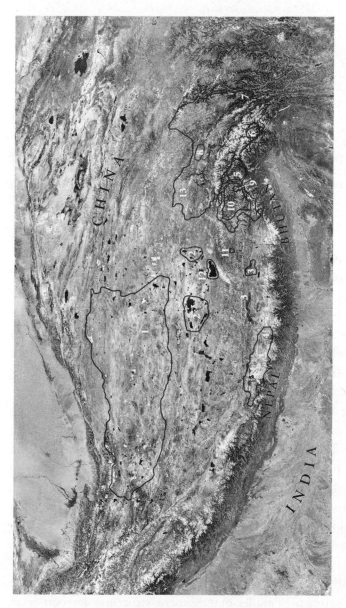

14.4 A Map of Thirteen Nature Preserves across the Tibet Autonomous Region/China that Followed from 'Discovering the Yeti'

Source: Future Generations University

National Parks
1. Changtang National Nature Reserve
2. Se-lin-cuo Black Neck Crane Breeding Ground
3. Yalung Tsangpo River Black Neck Crane Transient Area
4. Qomolangma (Mt Everest) National Nature Preserve
5. Yalung Tsangpo Great River Canyon National Nature Preserve
6. Chayu Cibagou National Nature Protected Area
7. Mangkang Snub-nosed Monkey National Nature Preserve

Region Level Preserves
8. Riwuchi Chang-maoling Red Deer Nature Preserve
9. Bajie Giant Cypress Nature Protected Area
10. Gongbo Regional Nature Reserve
11. Lhasa Lhalu Wetlands Nature Preserve
12. Namtso Wetland Nature Preserve

Integrated Protection Plan
13. Four Great Rivers Ecological Environment Protection Plan

Key to 14.4   Names of Thirteen Protected Areas across Tibet Autonomous Region/China

To reach here, we drove across the Tibetan Plateau, at an altitude where the oxygen is one-third of the way to the moon. Without trees and only scant grass, the landscape indeed looks moonlike. Homes are made from lifting up and packing the surrounding dirt. Four of us sit in an ancient caravanserai. Until the Chinese completed the motor road we rode, Shegar was a halting place for Tea Road caravans between China and India. Outside the gate, to get to India from this caravan stop, take a right; to go to Lhasa then China, take a left.

The just completed meeting in Lhasa got the support of Hu Jintao, Governor of the Tibet Autonomous Region (later to be President of the People's Republic of China). The decision that was made was rather than creating a separated management, protection for all four zones of the new national park will be through the local government. Parallel management structures for conservation and politics would not be set up; rather, one system will do both tasks so the result is people learning to live in an enduring way with nature.

Outside experts who have heard about our actions are sceptical: Can China's Communist Party truly be a partner? But China has an unexpected history of partnership. Barefoot doctors and peasant scholars were not the accepted models of how health and education could be delivered. In environmental protection can a people-based approach be used? Hu Jintao just told us the idea is to co-opt the political for the environmental, to change *to live with nature*. To me this meant developing bioresilience on the part of one species to accommodate to change: what will carry not only human life but all life is setting in place systems that grow our abilities to stretch.

John Craighead, Tom Roush (a friend from New York), and I sit around a big table, one of perhaps twenty in this cavernous hall, built for the Chinese army when, during the 1960s and 1970s, they feared attack by CIA-supported fighters of the Dalai Lama. With time people gain insights, and the Dalai Lama's enlightenment is proven by his present message of no longer advocating violent insurrection. In a world today tempted towards fundamentalisms, the journey of this man speaks: he began as a fundamentalist and took in new ideas. His Holiness stretched.

Across the table, Tom is talking: 'Dan'l, I know it sounds stupid, but that Everest we saw today has got to be the biggest thing in the world, and when in front of that mountain you know you're a long, long way from anywhere'. Tom is unwinding from the shock of discovering himself castaway atop a 17,700-foot pass when our vehicles abandoned him in the dark after our group stopped for sunset photographs. He asks, 'What prompted you to start this project, Dan'l?'

The talk has been of biology for the past hour, and that tired others in our group so they turned in for bed. 'Tom, I came searching for the Yeti. Some "thing" was making footprints in the snow. But what I discovered is a habitat; the larger context of the Yeti tied us to a quest for wildness's lost place in a today that takes away wildness'.

'If you want wildness,' John laughs softly, 'let me introduce you to *Ursus arctos horribilis*—you'll feel wild real fast as you scramble up a tree with a grizzly on your ass, leaving unrecognizable footprints on tree bark'.

'I know the story's Nepal side—the Yeti leading to Makalu–Barun National Park', Tom says. 'But how did you cross a closed international border, especially one into Tibet in China? You don't get opportunities such as we had in our just concluded meetings with three vice-governors when the border is closed. You cannot simply stop by their offices to propose an idea when you cannot even enter the territory where they work'.

'Work grew by partnerships, Tom', I answer. 'To get a partner you must offer that person something, and over the years what I've offered is information. In international work, Americans usually offer money or power. But having neither, I use information. When I became a friend with Nepal's Crown Prince in graduate school and then when he became king he needed information he was not getting from his own people, and I brought him that. His trust grew.

'But I want to know how you got into China? Tibet was totally closed', Tom asks. John is smiling, for he'd watched the story unfold over six years.

'As a climber, Tom, I learned that sometimes a crazy idea leads into an interesting climb. There are many routes up every mountain'.

'Tell about using the king's helicopter', John says with a grin.

So I start, 'When Makalu–Barun Park was underway, in 1985 an international conference was held in Nepal, and because our new park was based on partnerships with people, I was invited to speak. Attending that conference was a delegation from China. And it being an all-Asian conference with ministerial-level representation, the King of Nepal held a reception.

'My speech had been earlier in the day. I had talked about both sides of the border, extending from how Makalu–Barun was a core area in the pan-Himalayan wild, and I spoke of the role of the mountains in people's lives across Asia. At His Majesty's reception I went up to the Chinese delegation—they are very formal you know—but my speech had made me known, and asked the delegation's leader if, being from Beijing, he would like to look into his country from several passes between Nepal and China the following morning. He could see then what I had spoken of. I suggested we might arrange for the king's helicopter to show him. He was astonished, saying how could he, a minister from China, use the king's helicopter?

'Let's ask', I said, and we walked over to the king. I was greeted graciously, and permission was given. So the following morning we flew around the world's highest summits and peaked over into China. Three senior members of China's government saw valleys and mountains that called for being created into a national park. Introductions in Tibet followed.

Tom smiles, 'You could make a movie out of this'.

'That was only a door into Tibet—I kept gathering information', I continued. 'After the Everest preserve, I suggested we study the larger Tibetan region. The whole Tibetan region was incompletely known in the mid-1980s, even to the Chinese. It cost some money which I had to raise, but the results were staggering. Teams now of Chinese and Tibetan scholars have been going into valleys for five years. The original Everest park ended up connecting to five parks in Nepal and now on through Bhutan, India, Myanmar, into Sichuan in mainland China itself'.

Tom sits silently.

John has thought of another question. 'Dan'l do you remember that cave in the middle of the Barun Valley? You said you were going there someday to look for your Yetis. Dense jungle was at its doorstep, a stream running past, no human village for a week's walk. Nearby were grassy glens where baby Yetis could gambol in the sun. The climate was ideal, and food plentiful. Did you ever check it out?'

'Well yes. Last year I was in a helicopter surveying the boundaries for Makalu–Barun National Park. With our survey of the Ishwa Khola finished, as we headed to the Barun I asked to fly across the meadow where Derek spotted the cave. It was by chance the same time of the year, and the same time of the day. The chopper was 20 feet off the ground. Immediately that cave stood out—big, inviting, mysterious. I pointed the pilot towards it. Then, hovering in front, the cave disappeared. The black entrance so convincingly a cave turned into a shadow cast by the cliff above. At a 10:00 a.m. angle on a November morning, the sun creates the appearance of a cave.

'Ah, my last fantasy farewell', Tom exclaims. 'Lost Yeti survivors holding out in secluded caves, noble savages making love in the grass as their disappearing kind holds valiantly on to a balance with nature while other hominoids destroy it. Let's go to bed'.

❧

BUT THE BEAR ENIGMA CONTINUES TO PUZZLE. The tree bear and the ground bear, are these two bears or one? Science that answers mysteries by DNA and skull features says 'one bear'. The villagers of Shyakshila, who know their bears by behaviour, say 'two bears'.

In the villagers' definition of bears based on behaviour what matters is the likelihood one bear is more apt to invade their lives. If a fierce bear lives in their jungles, and another easily scares out of their cornfields, having two bears makes sense. But to organize life on our planet, science says both bears are one. A scientist can place fourteen skulls into a sequence and all differences explained within normal taxonomical variations. Nonetheless, skulls do not have behaviours—and concerning those skulls though once wild, what that dentition can do in fields and to people is a very different definition.

For science's questions, John's hypothesis is probably the finding: the tree bear is the juvenile, the ground bear is the adult, and likely the dramatic 'thumbs' are an inner digit in young bears still with flexible tendons where these digits are 'taught' to fall lower on the paw and to grip in a thumb-like way. A young twoling, that must spend much of its time in trees to avoid aggressive big males on the ground, trains its front paw to work like a primate's hand with opposable grip.

If so, what of the mystery of Oldham's *Selanarctos aboreous*, advanced a century earlier, a decade and a half before the Yeti was reported? In 1988, I spent a day in the British Museum, pulling out drawers of bear skulls. It was a delight to be in that unequalled museum. I was on the seventh floor, with bear skulls around, the world's greatest collection of Himalayan bears. I pulled *aboreus* out of a drawer where it had sat 120 years, and indeed there is nothing special about it. Pocock was right, and I am sure every taxonomist would agree: *arboreus* is another *thibetanus* skull.

I went to the library on the second floor and got Pocock's *Mammalia of British India*. He debunked *arboreus* on the basis of the skull, but described subspecies of *thibetanus* on the basis of coats, now an unacceptable taxonomic criterion. *Selanarctos thibetanus thibetanus*, from the eastern Himalaya, is smaller and has no undercoat. *Selanarctos thibetanus langier*, from the western Himalaya, has a thicker coat, and especially an undercoat. But in that article, in addition to the coat, Pocock described behavioural differences between *langier* and *thibetanus* of aggression and non-aggression, the differences we found between ground and tree bears.

What, I wondered, creates the behavioural differences. My conclusion now is that the Barun being isolated makes bears there less aggressive—as, in the reverse, black bears in the United States become more aggressive when they live close to humans, eat garbage, and no longer fear people. Bears when in competition for food become aggressive. However, Barun bears who can work the trees find plenty of food; they thrive as young bears and are not aggressive. But the big old bears, who as juveniles ate well, find less to eat on the ground as the dense tree cover minimizes ground level food growth. (Trees being the base of the life pyramid in the Barun, not grass.) These codgers get mean competing for food. Aggression then enables them to survive.

If that is the tree bear/ground bear explanation, what about the Yeti?

First, it is important again to dismiss all but the evidence that is genuinely mysterious as all alleged Yeti artefacts brought forward across the decades have been studied and explained. The most famous is the Kumjung scalp. Hillary's expedition in 1960 borrowed that and took it to the Field Museum of Natural History in Chicago. The 'Yeti skull cap' turned out to be a moulded portion of serow skin, probably from a serow's rump. In another monastery is a purported Yeti finger, but DNA shows that finger to be human. A Yeti bone exists in Bhutan, but DNA evidence shows it also to be human.

Hair and droppings collected by the Slick expeditions have never been made public. I have trouble imagining Slick did not have this done privately, and it interests me that towards the end of his life he was less adamant about the Yeti existing—maybe his privately held lab information was the reason. But as Slick never brought forward evidence, and all we know is the field reports of expeditions that went looking, the absence of evidence is evidence his reports should be discounted.

A report becomes truth only when objectively confirmed. And multiple reports of multiple sightings do not sum into evidence. Yet, the Yeti lives on. When a Westerner is on a Himalayan trek, the purpose is adventure, then that adventurer comes across a print in the snow, and he brings home the 'Abominable Snowman'. It is not evidence of the animal, but that this adventurer has been to the abode of Himalayan mysteries. In this manner, the Yeti is unlikely to ever die.

In this vein, I got a phone call from a Hollywood studio some years ago. As a 'Yeti expert' they wanted me to sit before cameras on a live TV screening as they played Yeti footage they had acquired. They would give me a thousand dollars. At the end of the showing I was to say whether the image was real ... or point to the zipper in the suit. I declined, so they found another expert. But Bob Fleming and I watched the footage when it showed. We did not see the zipper, but in the film's background Bob recognized a Scandinavian alder.

For the enduring Shipton print mystery that really started the Yeti search—and certainly my search—the suggestion of it being an overprint is, I am sure, the explanation. To understand this, it is helpful to look first at what the front and hind feet of *Ursus thibetanus* look like when the prints are made in the snow separately. The photograph 14.1 at the beginning of this chapter shows prints I found in the Barun Valley in 1986, with size indicated by a 52 mm lens cap. (Lens cap is 56 mm in exterior dimensions.) Noteworthy about the front foot is the three 'toes' on the upper left—these show strong similarities with the three toes on the Shipton print except that no nail marks are revealed. Noteworthy about the hind foot is the broad base. But what is most important about this rear foot is the nail mark on the upper left with a second nail mark beside it. The base of the foot shows strong similarities to the base of the Shipton print, but more crucially the two left nails on that rear print are identically placed to marks in the middle of the Shipton print.

The bear explanation for this print and the importance of the two marks in the middle of the Shipton print is not original to me. It was advanced by John R. Napier, a British primatologist who was involved in finding *Homo habilis*. Napier's Yeti argument is:

Something must have made the Shipton footprint. Like Mount Everest, it is there ... I would say a composite, made by a naked foot treading in the track of a foot wearing a leather moccasin ...

14.5   Jesse Washing Plaster Casts of Paw Print Moulds when We Placed the Hind Foot Offset to the Front to Replicate Various Configurations of Overprints. An Overprint that Comes Close to Repeating Shipton's 1951 Photograph Rests on the Ground by Jesse's Feet

Source: Author

The curious V-shaped kink behind the big toe, which has no apparent biological function, could be then explained in terms of a deep fold in the leather of the moccasin.[2]

Napier is saying that two people walked perfectly in each other's tracks.

Napier is right about it being an overprint, but not as a human composite. First, it's improbable that two people walking step after step in the anoxia of 19,000 feet would do exactly the same stride, even if they wanted to make the walk easier by stepping in each other's steps, and in this case having their bare toes always beyond the 'moccasin' print. (Additionally, as the snow, according to Shipton, was thin, little reason would cause a follower to seek to step in the front person's prints.)

Second, Napier is a primatologist, not an anthropologist. He did not know that in that part of the Himalaya villagers never wear soft-soled moccasins such that the bottom could develop 'a deep fold in the leather'. Traditional footgear such as that of North American Indians are in his mind. In 1951, before modern footwear came, a local person in the Himalaya would have worn either the hard-soled Tibetan felt boot or the rigid shoe of twisted grass rope. Both leave sharp indentations in the snow.

Other things are also wrong with Napier's argument. First, the supposed 'heel strike' suggested for the back of the print is in fact caused by melting, because it matches a similar indentation on the side (not back) of the partial second print shown in the bottom of the photo, an indentation almost certainly caused by melting. An important point, mentioned earlier, needs flagging here. Napier prudently went to the original negative of the Shipton print, and he found where the cropping of the photo had not shown the top of a second print on which additional inconsistent indentations are evident, especially with a nail-appearing imprint in the upper left, and which I attribute to be nail prints.

The variations between full print and partial print are important. They are consistent with a print being made by a four-footed animal that brings its feet straight down. Further, the Shipton print is concave, whereas if a bipedal hominoid had made them, the print should be convex. Bipedal walking requires an arch to launch the toes in each stride in their pivotal role. No arch shows in Shipton's print.

But the most determinative feature about the Shipton print, the feature indicating it was made by a bear, is what Napier called the 'curious V-shaped kink behind the big toe' referenced above. That is a nail mark; it is the front left nail of the hind foot. The hind end of the bear being heavier than the front, the rear foot pressed into the snow to show nails where the front foot did not depress to show the nails. Then on the right of this print is another less distinct nail mark that relates to the one on the left; the distance between the two perfectly fits the hind foot of the Asiatic black bear, precisely the right width separation for *Ursus thibetanus*.

Other clues in the Shipton account suggest a more complex explanation for this most famous of Yeti prints. The trail Shipton and Ward came upon was more than one bear making overprints. Recall (in Chapter Five) Shipton's quoting Sen Tensing having no doubt whatever that the creatures (for there had been at least two) that had made the tracks were 'Yetis'. Tracks of the 'at least two' were clearly confused. Here is my point: When does more than one bear travel with another? It is likely that Shipton and Ward were following a mother bear that had cubs walking after her, and this is what a second photograph in Shipton's 1951 book shows; this photograph (see 14.6), is of the whole trail.

The photograph is clear; one set of tracks is being overlaid on top of the other. It is not as Napier was suggesting humans trying to walk in each other's tracks, but rather young bear cubs dutifully following their mother. This photograph shows a hit-or-miss of footprints on top of others. Sometimes it is evident that the prints end up being big round saucers. Other times they appear like long-ish strides. (In the photograph also are Shipton and Ward's prints walking on both sides of the saucer-like bear overprints.)

This isn't the first time multiple bear prints were mistaken for the Yeti. In 1954, Sherpas claimed mysterious footprints to be the Yeti, but Charles Evans saw nail marks, and pointed out that one set was larger, also indicating a mother and a cub. Further, to support the mother and cub thesis, the season of Shipton and Ward's discovery is right for travel of a mother and her cub. If Shipton's Yeti is a bear, the explanation should fit both foot features and

14.6 Michael Ward Standing Beside the Trail of 'Yeti Tracks' That He and Eric Shipton Found on the Menlung Glacier in 1950—Note How the Trail Reveals 'for There Had Been at Least Two'

Source: Royal Geographical Society

known bear behaviour. Likely, the tracks that day in 1950 on the Menlung Glacier were a mother and cub(s) headed across a pass in search for food.

But, apart from an increasingly firm explanation to the iconic Shipton print, there is mathematical refutation against Heuvelmans's *Gigantoithecus* hypothesis of a race of giants, a side branch of *Homo sapiens*, that didn't die out half a million years ago, but survives as a remnant community in remote Himalayan valleys. For this, John Craighead astutely argues about minimum viable populations, and I earlier referenced his argument. Minimum viable population mathematics pretty well debunks the Yeti as a hominoid.

Without data on Yeti reproduction, let us start with Craighead's grizzly numbers. Thirty individuals is the smallest population of grizzlies that can survive. Assuming hominoid intelligence, a Yeti might require half that number. But as hominoids spend more time raising their young than the three years of a grizzly, and the reproductive period is longer, plus the likelihood of maternal death is probably higher, these variables raise the minimum population. However, arguing from the other position, the Barun and its neighbouring valley in Tibet is ideal, isolated, food-rich habitat, so it might be possible that the lowest possible number of Yetis could be twenty individuals. For the Yeti, therefore, what is needed is a minimum population of between twenty and thirty individuals.

That number also fits with humans where the settlement of the Pacific gives evidence on minimum viable populations. Anthropologists have shown that a human boy and girl going out in a canoe to settle a new island will unlikely be able to reproduce beyond several generations. Larger concentrations are required of twenty or more even in the idyllic, food-rich, disease limited, Pacific Islands. A boy and girl, lying on a beach having sex eating coconuts and fresh fish, will happily make babies. But over several generations (aside from inbreeding) they are unlikely to create a community.

To put it another way, once the population of an animal that must live by hiding gets so low that the population is impossible to find, the population is probably so small it is gone. Or, to put this in terms of the Yeti search, since a population—not an individual—must exist, the search is not for a solitary animal. And after a hundred years of searching, if not even one animal has been found when at least twenty must exist, the population is nonexistent. Statistically speaking, there is probably not even one Yeti.

If, then, it takes a village to grow a people—how did humanity or gorillas or bears (or chickens from eggs) ever start? Twenty new babies are not born at one time. (Adam and Eve, by this proof, are mathematically not enough to start the human species.) Evolution is the explanation. A genetic mutation occurs, but not such a sig-nificant mutation that the new being cannot breed back to the prior species. Closely related species can breed to each other. (Consider all the new DNA evidence to show that humans were breeding with earlier hominoid species. As a separate species, humans evolved to be distinctive over generations.) Speciation comes from breeding back, and then numbers growing as speciation differentiation con-tinues. A new species is not formed in the first mutation. Multiple generations grow the minimum viable populations, and in these years ongoing mutation reinforces separate identity.

More on DNA's role solving the Yeti enigma is worth mentioning. More definitively than skulls, DNA now identifies species. So, it is essential to subject all alleged Yeti relics to DNA analysis—a new animal might just turn up. With this objective, the Oxford professor, Bryan Sykes, led a team that systematically examined all 'anomalous primate' artefacts they could recover. That explained all but two based on known animal DNA, and the two they advanced might connect to a remnant population of bears in the Himalaya.[3] This possibility again ignited Yeti speculation—in 2014, the Yeti was back. A subsequent study by Eliecer Gutierrez and Ronald Pine repeated this analysis. Working with parallel mitochondrial 12SrRNA sequencing, they showed that the short genetic sequence was not adequate to support a conclusion of an 'anomaly' and was almost certainly from the Himalayan brown bear.[4]

Lore of this Yeti mystery began in the 1890s, continuing through the 1960s when in the Western view there still existed 'undiscov-ered' Himalayan valleys. Discoveries were happening still then in the Himalaya. The great summits were being 'conquered', and the would-be conquerors were finding the footprints. Today with the Barun explored—and not even Lendoop finding the Yeti (which is hugely telling)—the remaining area of sufficient size where a Yeti might live is Tibet.

In my now three decades of conservation work starting nature preserves, I've been privileged to work in every prefecture in the

Tibet Autonomous Region. I've gathered Yeti-related evidence at every opportunity in forty-five exploratory trips. Yeti stories abound—from Tibetans in the west along the Tsangpo River and also a thousand kilometres away in the east at the headwaters of the Mekong. They speak of *Dremu* who steels sheep, an animal that has human footprints. They speak of *Metoh Kangmi* that travels alone, has ears flopped forward, lives high on mountain ridges, and can walk erect.

Careful probing of these reports give details pointing to a known animal, once again a bear. Some Tibetans have actually seen the Dremu and do not pass on someone else's story, a large head compared to the body, long teeth, yellow hair around the face, long snout where the nose and mouth seem one. For years I was puzzled. Then one day at the Lhasa zoo I saw two animals and an explanation connected. The Dremu is the super-rare, never-studied Himalayan blue bear, *Ursus arctos pruinosis* (or sometimes a subspecies labelled *isabelensis*).

That identification connected because decades before, in 1961 as a sixteen-year-old, I asked the former king of Bhutan about the Yeti (the Bhutanese now call this king The Third King as his grandson, The Fifth King, sits on the throne). The Third King was an outdoorsman unusually knowledgeable about wildlife. He smiled when I asked about the Yeti, and it was clear he had given the matter some thought. He said he thought that the Yeti was a blue bear that sometimes crossed into Bhutan; of all the world's bears, these live at the highest elevations. He went on to say he was certain, as king, if actual wild people were in his mountains, he would have other reasons to know of such.[5]

In 1961, I was not happy to hear that the Yeti was a bear—and from the king of most pristine part of the Himalaya, a wildlife expert who also knew his kingdom's secrets. But now his conclusion is comforting. In 1961, Yeti searching was at a fever pitch in Nepal; it would only be logical that the king next door who had extensive jungles and snows would wonder: Is the Yeti here? And, it was clear as The Third King and I talked, that His Majesty had worked through all the options.

I had never heard of the blue bear. It was not until I saw that strange bear in the Lhasa zoo that I started to read up on this extremely rare animal. Then twice later doing fieldwork on the Changtang Plateau I saw individuals of this bear, once a male alone and another time a mother and two cubs. *Ursus arctos pruinosis* has huge shoulders and head, disproportionately large compared to other bears. I have not tranquilized this animal and mimicked its footprints in plaster as I have done with *Ursus arctos thibetanus*—perhaps there is no need for no one has yet brought forward photographs of Yeti footprints from Tibet that need to be matched—but given what we know about Yeti behaviour (and I saw the blue bear walk bow-legged out in the wild) and the paws I studied through the cage in Lhasa zoo, I am certain this bear, like all bears, will make overprints and that when going uphill, those overprints will elongate and look bipedal and human-like.

Notes

1. Wendell Berry, *Home Economics* (San Francisco: North Point Press, 1987), pp. 138, 143.
2. John Napier, *Bigfoot: The Yeti and Sasquatch in Myth and Reality* (New York: E.P. Dutton, 1973), p. 141.
3. Bryan C. Sykes, Rhettman A. Mullis, Christophe Hagenmuller, Terry W. Melton, and Michel Satori, 'Genetic analysis of hair samples attributed to yeti, bigfoot and other anomalous primates', in *Proceedings of the Royal Society*, B281: 20140161, recovered 26 February 2017: http://rspb.royalsocietypublishing.org/content/royprsb/281/1789/20140161.full.pdf.
4. Eliecer E. Gutierrez and Ronald H. Pine, 'No Need to Replace an "Anomalous" primate (Primates) with an "anomalous" bear (Carnivora Ursidae)', published in *Zoo Keys* 487:141–54; 15 March 2015.

5. Modern Bhutanese Yeti reports, of which there are many, are not
   dealt with in this book. They fit with Nepali and Tibetan reports
   with features like the backward feet—and Bhutan has high mountain and jungle habitat of significant expanse. In 1961, The Third
   King did not suggest that the Yeti was a real animal; his curiosity
   about possibly a real animal came from the Nepali footprints, but
   he did speak of a spiritual animal his people believed in. However, it
   seems in the late 1960s into the 1990s, Yeti sightings as animal were
   reported in Bhutan, but now Yeti sightings appear to be fewer. A
   recent BBC report studied why, concluding that Yetis are less discovered because children spend less time in the mountains, and in so
   doing have less opportunity to find mysterious footprints. Relevant
   about this explanation is how Yeti footprints are being found by
   children who then live their lives with this childhood mystery. As
   with me! The footprints were not being found by experienced people
   of the mountains, like Lendoop, who could identify animal signs.
   Also relevant is that the king never believed these stories. Recovered
   1 November 2015: www.bbc.com/news/magazine-3448314.

fifteen

Discovery

THEORY

Discovery

Previous page:

15.1   Petang Ringmo Camp—High Pastoral Camp to Which Herders Come on Makalu's North Side

Source: Author

October 1991. In the rock above, crouches a snow leopard; it has crouched there for tens of thousands of years. I am in the upper Barun Valley. Pilgrims who descend this path I have just walked believe the leopard guards Lord Shiva's sacred cave just around the corner. For geologists the leopard is a chemical stain in the cliff. To mountaineers coming to Makalu, the leopard portends avalanches unpredictably pouncing onto climbers.

Looking at the snow leopard, I wonder whether Rodney Jackson is right. Rod 'knows' snow leopards; he was the first to tranquilize and radio track this animal, and then spent three decades studying it. And what we have learnt is that snow leopards are more than just an animal—certainly the real animal's mating yowl, which reverberates through valleys from January to March, which people have compared to a horse, could explain the call attributed to the Yeti. But the snow leopard's call has also become a plea for the wild Himalaya. Where the Yeti speaks of a mythic wild, photographs of the snow leopard speak of the danger to the wild.

The physical body is one way to describe a person, so are their ideas, and why not also their connections. Trails walked in wild cities differ from trails walked in wild nature ... but how? The earth herself shifts, as do seemingly solid mountains when they quake. The habitat of living is alive; its constant growing that we cannot control is the essence of the wild.

On the edge of extinction until a few years ago, snow leopard numbers now come back, so much so that the species is taking a 'tax' in sheep and goats from its sister species, humans, where herders and their flocks also live in this animal's home range. Whose home is primary? Because of the Qomolangma (Mount Everest) National Nature Preserve in Tibet, snow leopard numbers have rebounded so much that animals left that preserve and crossed here into the Makalu–Barun National Park and the QNNP's western fringe, the Annapurna Conservation Area. In its recovery, the snow leopard shows how human life with wildness can return.

Snow leopards are like tigers: solitary from the time they leave their mothers. As adults, they move through their home range alone. But unlike tigers, usually two or more snow leopards move simultaneously through a home range, signalling their presence to one another with scratch marks they scent at rock outcrops on their routes as they pursue their primary prey: the blue sheep. Travelling at dawn and dusk, then stealthily waiting to pounce during the day, sometimes one snow leopard follows another down the same route, usually never closer than a mile apart—except when they join to procreate—the male then leaves, and three months later the mother typically has two to three trailing cubs.

Bears with two identities. Animals crossing the land making overprints. Snow leopards making ethereal yowlings. People filling in the details. These are all about the animal—but what about the habitat, the wild. For this trip I've returned to work through a hypothesis of the Yeti's changing portrait. We are the most dangerous of species—pressing the wild into remnant pockets—in exterminating the way life used to relate, we created a new age. On this trip, I have come to see if I can walk into the old pristine wild.

In a knee-wearying descent today, I came off Shipton's Pass on the southern ridge of the Barun to reach this series of meadows that cascade like ponds of green, a place called Yangle Karka. Most foreigners at this juncture go up-valley to Makalu Base Camp, seeking ice, glacier, and lofty summits. A few Japanese botanists as well as McNeely and Cronin came and went down. That is what I am going to do. The descending meadows of Yangle Karka are, the shepherds say, the last good grazing before the dense jungle of the middle Barun.

Every year for 200 years (maybe hundreds before that), herders have come. Like most Nepalis they're social, preferring village life, missing families and neighbours. They come, but they're afraid in coming because a snowfall might trap them here. They come for one thing, grass, as much as their animals can carry home in bellies and bodies, avoiding going lower into the jungle afraid of its animals and spirits. To expand the grass areas, on those weeks when this upper valley is dry, these pastoralists burn the slopes to push back the bushes.

Grass is also the reason why I've come. Studying the grass may give me a clearer understanding of the balance of wild and domes-ticated nature. In the people's fear is my hope to find a pristine balance that the life of shepherds has not touched. I seek a belt of wildness between the grazing of the upper valley and the pen-etration into the Lower Barun by the villages of Shyakshila and Shebrung. Lendoop says he has never even entered this Middle Barun where a human presence rarely passes.

I will be walking alone. I seek to listen for a wild humanity in me—and through that unlock an understanding of myself. The jungle of firewood, medicinal plants, and animals to which people come are a human supply closet. All the summits around have been trod on, those are the end limits of people from the city. But I seek places beyond the end limits of pastoral living or human conquest. If such balance exists anywhere in the Himalaya, it may survive down-valley ahead of me.

Here at Yangle Karka, year after year more and more domestic animals come, and one signature of this, beyond burning back the rhododendron and juniper, is the changing ratio of grasses and sedges. Jungle ungulates (musk deer thar, serow, and ghoral) don't alter the balance in their foraging. These animals, watchful for leopards, do not stand in the open. But domesticated animals, protected by herders, eat and trample in one place, and as grasses diminish, the more fibrous sedges take root. I seek a meadow at least three-quarters grass, where tiny buds on green stems display the pure wildness.

Yangle Karka's green pond-like meadows irregularly descend down the valley. Amid rocks in one meadow, I find the highest flowering plant in the world, *Stellera procumbens*, growing in a tight white cushion, having adapted to the altitude by condensing its blossoms to protect itself against the cold. Mountaineers climbing Makalu found this flower with its five deeply divided petals 9,000 feet above these rocks. Here it is low for the plant. At the lowest meadow is *nigalo*, the thin bamboo, its shoots the highest naturally occurring red panda food. In the rhododendron beyond is *usnea*, the hair-like lichen favoured by musk deer. So, with this food available, jungle animals will come this high. Somewhere from these meadows leaves the trail used by smugglers heading into Tibet.

಄

WHEN DID HUMANS START RESHAPING THE WILD? Active reshaping began two millennia ago. China cut the forests of Sichuan for iron smelters while Rome deforested the Mediterranean. But global reshaping arose with the British Empire burning coal to drive England's belching industries. The climate began to be remanufactured. Jesse Oak whose early steps were in these Barun jungles became fascinated by that.

Jesse identifies the moment in February 1884 when John Ruskin 'ascended the lectern at the London Institution and brought to his audience's attention "a series of cloud phenomena, peculiar to our own times ... which have not hitherto received any notice from meteorologists," a plague-wind darkening the skies across the British Isles and indeed all of Europe'.[1]

Jesse notes how Ruskin, with literary sources linked to paintings of clouds and sunsets, connected earth's climate being 'remanufactured' resulting in Britain's loss of natural richness. Ruskin as an art historian had realized 'the inextricability of the two',[2] linking a changing planet's climate and 'a society worshipful of what Ruskin elsewhere called "the Goddess of Getting On," ... and that such changes heralded a much deeper contradiction within industrial modernity itself'.[3] That changing of nature that would alter the world's climate had earlier been girdling earth's subtropics.

The environment in which we live is reshaped by the way we live. Reshaping occurs not because we so intended—but from living in unintended consequences radiating out from the ways of our living. We have moved from 'where meaning in the skies once bespoke the will of God, it is now bequeathed by the effluence of human affluence'.[4] We are awakening to the reality that we have reshaped the planet itself into an unintended consequence. Jesse Oak, a child who began in these jungles wondering about heffalump traps and 'isn't it funny how bears love honey' now as a professor seeks to find meaning in 'the effluence of human affluence'.

The term for this is 'abnatural'—an idea that inhabits the concept gap between the natural and unnatural, accents nature as an idea instead of its usual physical essence.[5] More than the identity as a real animal, the Yeti lives also in this space between the idea and the physical—placing ourselves with the wild. The Yeti is neither fully present in nature nor fully absent. Like a novel whose words may be fiction, it tells a story of true meaning. The human quest, as we step ever further from the old wild, seeks to bind nature and life experience. Doing so creates the *experience* of living. Our lives become worth living when connected to deep roots.

∽

ROLLING UP THE NYLON TARP I sleep under, I say farewell to the large fir above. Yesterday I discovered grasses here comprise less than 50 per cent of the groundcover. So Yangle Karka is heavily grazed. The splay of this fir's branches suggests its grain to be exceptionally straight. The next human impact here might be the tree. *Abies spectabilis* is the only Himalayan tree that easily splits into boards with hand wedges and sledges. One summer, herders as they graze this meadow will fell this tree, spend their days as their animals eat the grass, splitting the tree into door and window frames. At the end of the grazing season, they will walk out with the boards carried by their animals. Ten per cent of this tree will typically move, the rest wasted in hand cutting and splitting.

And after trees start coming down here, will houses go up? Maybe not if the Makalu–Barun National Park management works. Maybe this upper Barun will move slowly and non-threateningly not just from its probable future of becoming a village to a new possibility of being a core zone of the park. There will be no villages then, but more outside visitors to spend time in 'the wild'.

Such change may connect us as people to earth's forces. Chinese call this energy *Qi*; native Americans call it *Medicine*; the Hindus, into whose Ganges waters the stream at my feet flows, name the force *Brahma*; the progressives back home call it *Gaia*. These are forces greater than *conservation*, speaking of fires extending from the inner being and to the interface of our interacting with the great beyond. Their intent is to regulate individual greed.

The path will be leaps and jerks, as I was reminded when descending from Shipton's Pass. In descending I was reminded that life changed not in steady regular gradients. I did not feel the slight rising temperature of 1°F every 300 feet, but still at one point I moved from walking in grass to walking through bushes. Nature showed me what I could not feel. Continuing down amid juniper and rhododendron, where I am now I walk between trees. Grasses did not intermingle with bushes; there was a line. Nature changes with punctuation marks. I am in forest now.

Qi, Medicine, Brahma, Gaia, ways of believing, in their philosophies also speak of moments when life breaks through. Daily actions are underway, seemingly serene—perhaps hypnotically so, thinking little is happening like descending through an alpine meadow. Then life, which appears continual, acts with exclamation. I walk alone so I can understand the species whose actions create the new wild.

It is not just life systems on the planet that get punctuated, but the shape of earth as well. Reshaping of the earth stands before me. Ahead the whole valley changes. Behind, where drains the 12 square miles of the Barun Glacier, the valley is spread out and the river curls through meadows. Ahead, the valley restructures from U- to V-shaped, as a river once-caressing through meadows surges, its wavetops flecked white as courses and carves the valley.

The river narrows to a 15-foot-wide torrent that bounces from the south wall, blocking my further descent. If this is a barrier for me, it may have stopped others, especially herders with flocks. The opposite side of this moving fence might be pristine. The depth looks knee height, but I cannot risk being swept away like Lendoop's daughter. I will use the fast-moving river to sweep me across. So I remove a few key items from my pack, tying those to me, then the pack's contents are double-wrapped in plastic, and restowed. Uncoiling a light nylon rope, I loop it around a tree, tying my pack to the middle.

Wearing a T-shirt to cut the wind, socks that will grip slippery rocks better than bare feet, I fasten a sling with a metal snap-link around my waist, wrap the doubled ropes through the snap-link, and back into glacial cold water that slams onto my legs. Leaning against the rope, the water's force angles me against the bottom and I turn my body diagonally like a rudder on a boat, the force of the water pushing me as my feet shuffle along the bottom. The current drives me across like a pendulum.

On the other bank I sink to a quiet space above ground, out of the wind, absorbing heat from sun. A griffon overhead turns. Its hopes for my future are mistaken as I flex my legs, pumping blood down the arteries. Untying the figure eight at the end of the rope, I pull on the line, its middle tied to the pack. As the pack hits the water, it charges downstream. Ten pulls, and soon I pull a candy bar from inside the pack.

One joy of life in the mountains is the time spent regaining our strength and looking at birds move on mountain waves. Currents of nature give movement to our lives. An oak rises above. What is the difference between this oak grown from a fallen acorn that a rodent missed and an oak that would grow if I picked up an acorn here and planted it in my yard in the mountains of West Virginia? Is an oak an oak simply because it grows from an acorn? Is the snow leopard in the cliff behind more alive than the snow leopards in the Seattle zoo? Sprung from similar seeds, when does the wild become a garden?

Thoreau believed that in wildness was the preservation of the world. He might be wrong for people today; for this was an opinion, but also fact based on the life he led. In any case, the world he craved for is no longer with us. That type of preservation is at best now in gardens that have walls around them. In this trip, and perhaps in my crossing of the river, I have come to stand apart from a world of human control. I have come to touch the wild as when it was, as with Thoreau, the entire world, not remnants.

Descending, later I enter a meadow, a clearing I suspect made by an avalanche off of Makalu above. Maples, magnolias, and birches surround, much like those in the zone above Makalu Jungli Hot'l. But unlike Makalu Jungli Hot'l, which is 3,000 feet lower and grassy because of grazing, what made this meadow? Why do no trees grow? Each winter may pound a deluge of ice off the cliff above; perhaps a chute in the rock above channels avalanches that flatten everything that starts to grow except grass that comes back each season as the snow melts.

However formed, the meadow through my hand lens shows that grasses predominate; sedges are maybe 15 per cent. In waves of grass, I have come to an arboretum attended by primeval biology. In this meadow is a place for me to listen inward as well as outward; here, I stand equidistant between unpeopled DNA and the stars, the world before people began changing the planet. At the base of a large oak I prop my backpack against its wizened trunk, two rocks are rolled beside, then my sleeping pad wedged between: a seat with armrests. The clearing is no longer as it was. Tomorrow when walking by the river's edge looking for prints in the sand, I will leave mine.

I have come for the wild, and in doing so my first actions are domestication. In the millennia before fact-based thinking advanced results-based formulae of the scientific method, human knowing presumed an integrity from the great beyond. Truth came from mystery. Now the assumption is that the method is the way to lead to knowledge. But that does not explain the Brahmin schoolteacher, the villagers of Shyakshila, the recognition of our daughter, or Grandpa and the tigress. The belief that truth comes only by verifiable fact directs people to seek in tunnel vision, crediting they see distance using a telescope, then turn to a microscope trusting they see deeply. Tubes of our making certainly delineate into the beyond, but they also block empathy informed by unknowing.

Himalayan sages speak of secret understandings. These are more than symbols of portals to an unknown enlightenment—they speak of more just as a rocket ship is more than a symbol of a physical voyage beyond the planet, or five interlinked Olympic circles beyond our quest to the limits of the body. In the symbolism of such a calling, the Yeti too is real, embodying from which we have come.

15.2 A Photograph of Two Yaks and Their Herders in the Valley Just North of the Barun in the Qomolangma (Mount Everest) National Nature Preserve

Source: Author

I have lectured at the Royal Geographic Society in London and its Hong Kong audience explaining the Yeti tracks. (This society being the group more than any other that gave the Yeti credibility.) In America, I have presented bears skulls on television (the medium of that nation) and shown how animals with such skulls make footprints that look human. Nonetheless, people still call for a Yeti. The relative people seek is tangibility to human identity. Riding this symbol, life becomes redefining the mysterious wild into a human-faced image while remaining wild inside our souls.

BEYOND THE OAK IN THIS GLEN STANDS A PLANT one-metre tall with dark green leaves and a flower: the may apple of the Himalaya, *Podophyllum emodi*. The altitude here is low for this increasingly rare plant, but the habitat is appropriately wet. Village plant hunters stalk this, seeking the egg-shaped, purple, pulpy, heavily seeded fruit to sell to pharmaceutical companies. A commercial voyage of sale to cure cancer begins by pulling up these apples by their roots. If I'm finding this plant, I am now beyond the travel of village people, because if people came to the place, this plant also would surely have been pulled.

My eyes probe for other signatures, perhaps the more valuable *Rauwolfia*. Thirty years of hunting has almost exterminated *Rauwolfia serpentina*, a red-flowered, coffee-like shrub whose root cures hypertension. At the edge of the meadow, I find *Nardostachys jatamansi*. A curious little plant, long, tightly furled leaves binding miniature 'maize ears' on a stalk; its roots have a fragrant oil, an apparent cancer cure. Again, to harvest this, the plant is uprooted. If *Podophyllum emodi* and *Nardostachys jatamansi* are present, there is no need to study the grasses.

Hunger for cures from the wild circles the world. A quarter of medicines sold in the US originate in plants. Hunting intensifies as potential identifies other plants from the wild which might cure the ills of modern life. Nepalis, pushed by population growth to work marginal land rather than the no longer available valley bottoms have been recruited to seek these alternative crops that they can lucratively harvest but do not need to plant. And, recognizing the pressure on these resources, conservation groups try to save the forests here and around the world. In that, their argument curiously flips: protect forests so more plants can be taken out. Dad and I did this for the *nusha bhoota* whose aroma puts passers-by to sleep; that was my first 'big' Yeti expedition.

Humans depend on plants which feed, heal, and house us; our planet takes colour from their diversity, and their luxuriance consumes our wasted gases. Plants are at the base of the pyramid of life. And the species at the top, despite religious teaching, never stops wanting ever more. On this planet of human manufacture, I have arrived in one meadow from which people have not come to take.

Beyond the *Nardostachys jatamansi* is the jack-in-the-pulpit, *Arisaema constatum*. My childhood image of this plant was a cobra with an emerald hood hovering over a coral throat. For centuries, Tibetans descended into these jungles seeking this tuber that looks like a yam and if eaten raw makes the throat swell and the eater die. They learned, though, to grate the tubers, wash the shredded flesh to remove the soapy saponin, and then ferment the mash to separate out the toxin. But like so many products from nature that we must detoxify to bring into human life, this mush must be washed again, spread thin to dry, then ground into flour, and then, astonishingly, made into bread. How did the Tibetans figure out such chemistry, as getting flour from barley or wheat is much easier? How many people experimented with eating, choking, and maybe dying? More improbably, how was the discovery made that this flour effectively cured piles?

Jack-in-the-pulpits impart metaphorical lessons too. As with the leech, this organism is both male and female, dual-gendered but not at the same time. A young Jack, if it flowers and produces a lot of starch, gathers sufficient strength to turn female. A 'He' becomes a 'Jill'. Yet Jack cannot mate with itself, for the male flower has died by the time the female matures. So, a gnat slides up and down the soapy male stalk gathering pollen, slips out a hole in the bottom of the cup, flitting on (hopefully) to a female flower. Entering, drawn by her fertility, the gnat slides up and down in her stalk, but in the female exists no bottom hole, and so the insect dies contributing pollen and its now-rotting tissue to the female energy.

If I prove to a disinterested world that the Yeti does not exist, that its footprints are made by a bear, and its voice that of a snow leopard, what animation will replace to speak for an increasingly limited wild? If the Yeti travels to a nonexistent certainhood because we know it not as a wild hominoid, do humans lose the ability to believe in the wild?

Fortunately, other ways exist of coming close with the wild. Whenever possible, I camp with a fire. Fire is humanity's most ancient domestic possession, a feature we've carried across the ages, a source of fellowship that brings us together, where we gather for so many types of nourishment. Burning fires turn humans to believe they are sages. As night falls in this unpeopled glen, I collect fallen branches.

My fire method places two large sticks parallel, then hair-like twigs across that will soon catch; above them larger twigs so rising flames feed crackled sticks. Some people now turn to fire starters, but in so doing they break with a skill perhaps humanity's oldest, passed across hundreds of generations, connecting today to ancient roots. As a youngster, I carried flint and steel and dry moss, items I traded for from a Tibetan refugee who recently fled into the Himalaya from Tibet. And from hunters I learned which leaves were more water-repellent and might shield dry twigs underneath on a day of rain. Tonight, a match ignites the hair-like twigs.

Flames lap thin stems, releasing life long-passed. Energy is unleashed that came from the sun to the earth years before. Fire can be viewed as the reincarnation of the once dead, as vibrations that journeyed across the solar system again pulsate. As hominoids stepped from wildness forward to domestication with technology, arguably a first step (after using a walking stick) was to gather three stones, walls beside the flames on which to set food—then from the three stones, homes went up, hearths were laid, and cities spread. Humanity's departure from wildness began with harnessing fire.

The fire before me focuses ideas, helping them dance inside. In the dark heavens above burn stars, taking those ideas to fires light-years away. Two fires grasping inward and out, eyes and I, both see but do not know. As the eye carries over great distances, the mind similarly pierces inwardly, sight and self, adapting from the intensity of the burning to the vastness of the dark around, seeking what is hidden.

To what extent do I have a home in this Himalaya? I grew up here, but am not a native. In coming here, I have walked into its reaches; an alien of different-coloured skin, riding cross-generational questions. A century before our family came as medical doctors giving compassionate care—they also came to evangelize a faith not indigenous to these mountains. I am a new evangelist bringing the idea of empowered people's owning their future.

In my mission I come with assets that few in these mountains have. Though I do not have financial wealth, I can call for a helicopter to lift me out. I can call on the highest leaders in both Nepal and China to work as partners. I can call on banks to lend me money. Although not taking from the place to enrich my life back home, I must ask whether my coming to help grow a national park might be a new colonialism?

Grandpa and Grandma were part of the old colonial. Yes, Americans, not British; yes, missionaries not merchants. The family was not traditionally taking, but because of birth and learning, they gained access to the good life in India—bungalows, servants, first waiting line positions. Against that pressure, to guide the family towards the giving, we had Grandma—she pre-eminently cared. All knew that, certainly I her eldest grandson, for to be a male in India cloaks one with privilege as does caste. Grandma opened understanding that pressed me to go beyond my privilege. She welcomed into our lives hundreds of children of lepers that overflowed her house, she stepped to her veranda sometimes hourly to give medical care. She taught me that touching the outcast brought me in.

I have come here now to touch that which once as a child I knew. Night being still young, I step to the edge of the glen to bring more dry branches. My eyesight stops against the valley's far wall. Partway up, a light shines. Might it be the moon off mica-filled rock? I step sideways to see if the light disappears. No.

From deep in my soul is recognition—only two things make fires, humans, and lightning.

Tonight the sky is clear. I am not alone. Is that fire smugglers? Or is it a fire made by some other? Might it be …? When daylight comes, without its beam, I may have difficulty being guided to its hearth. Rolling up my sleeping bag and putting all into my pack, I throw a rope over a branch and hoist my supplies from chewing animals. Into the dark I step.

The night's challenge is figuring out the route when walking. Steps are easy, a flashlight shows that far. But the beam does not reveal the trail. Lit is a tunnelled path showing more what is left out, causing the dark to feel even bigger. As I walk towards the distant fire, to prevent my approach from being seen, my hand covers the lens, separating my fingers to let light sliver through. On that distant slope the reddish light continues to glow. It's likely not herders; this is not the season. Plant hunters? Possibly. It could also be the Yeti … ridiculous, even if I am in its *sanctum sanctorum*. Most likely, smugglers. Approaching a smugglers' camp at night....

This most remote part of the Barun is where Yetis would be, if they've survived. And if Yetis exist and if they've survived, why could they not know how to make a fire? Doing so here would be safe, for they know what the plants have also now shown me: humans do not come here. Having a fire speaks of an animal that gathers with its kin—eats cooked meat. Having a fire says that the Yeti, if it lives, does so with a population.

With fire to fill their eyes, those sitting by it will be less likely to see my tiny light as fingers open a larger slit when I must see more, then narrow to almost nothing when steps are routine, always making sure never to point the lens to that slope. In the heavens above a satellite circles. On this earth today, not just here but in a few other remaining jungles, it is possible to walk towards pre-humanity while satellites circle above.

I try to be quiet, then laugh, there's no way they'll hear me from here. It was the answer of moments ago that is the most logical …

and smugglers do not appreciate being walked up on in the dark. Cautioned, I turn back to camp.

Before crawling into my sleeping bag I notch three sticks, tie them in the centre, and splay them out into a tripod. A fourth stick is aimed carefully through the fork like a rifle to point the direction in the morning. In the warmth of my sleeping bag I wonder what would I do if I met the Yeti. A first need, if I ever meet the Yeti, must be to communicate no danger to this animal is that has survived by not communicating.

What should I say, hominoid-to-hominoid, if I just happen to come upon a Yeti, and it seemed possible I could communicate. It is an intrigue I've pondered over the years. Walking upon a tiger and a bear I know not to shout. But with the Yeti what is the action? Maybe speaking Nepali or Tibetan should be a first attempt, or should my sounds be simply some tone? How about: 'I've been following your footprints for thirty years, quite an elusive trail you've left'.

Would it talk back and say to me: 'That's strange, because your imprints on the land changed my world forever'.

Should I pull out of my pocket the language of food?

People are afraid of the Yeti. But why? There is no evidence that it is dangerous—and so no reason to fear. Like other primates, might the Yeti even have a sense of humour and reply to my question, 'We beasts are the apparition humans see when you link an unsuccessful mountaineer with a giant ego'.

Human brains have been evolving, that aspect that allowed us to become sapiens and wise, kept coming up with ideas. Evolution was underway, but until language arrived, nothing showed collective wisdom of those ideas. Geniuses will have existed. But then something happened to us as a species. Being able to communicate to a Yeti might shed some understanding. For in our evolution suddenly a lot of evidence appears of human minds at work: tools, houses, agriculture. Language must have been what changed a solitary genius. Language is about connection, like dancing fingers of ideas, the expression of people controlling the processes of living. From my sleeping bag I look at the distant light above.

If the Yeti has eluded for so long, it must have communication—at least among its kind, if not also understanding us. It has to have had a language beyond yowls across the valleys. It therefore cannot be a prehistoric hominoid, certainly not *Gigantopithecus*, which separated from Homo sapiens a million years ago and had weak language; probably it is not even Neanderthal or Cro-Magnon. A Yeti successful in its hiding must be one of us, or much like us. In the fire beyond proof may be glowing that I am among my own kind, whatever our group might be.

The morning shows night was not a dream. The tripod stick points to a diminutive cave that to my binoculars appears empty. It takes two hours to reach the ledge on which it sits, maybe 300 feet above the valley floor. A screen of bushes partially hides the opening on a shelf of up-sloping ground. On it are three old fire scars. In each are loose ashes. But from one smoke thinly spirals. As with bipedal walking, fire signatures what only our kind makes.

The smoke's rising filigree loosens as the spiral expands; hair-like wisps float apart, disappearing at the roof of the ledge. Is this fire's maker near? I look to the ash, for ashes carry traces of our ancestors' fires across the ages. Ash can lay inert for millennia, holding its story of once providing warmth and maybe food. (It was carbon dating of ashes found in Barun soil that proved people have been in the Barun for more than two centuries.) But the story told here with the rising filigree is more than ashes.

I remember other smoke curling in the Barun—when John Craighead and I were at Saldima, and it was rising beyond the waterfall. We wondered whether those wisps were mist from the hidden lake. Then my eyes see a different residue: a footprint in this cave. My hand touches it; it is human and huge.

The Yeti is a wild animal. I have proven it a bear. Might this print that is clearly human be made by a people who have cut themselves off from society? Such recluses could, as they needed, rejoin with people, bringing harvests, going to Khandbari to sell and buy. Living in the wilds their feet would become splayed like the print before me, capable of crossing high passes and walking snowfields. The message before me speaks in duality: In this shelter, last night both fire and footprint were signed.

Outside are no other signs. I could remain here and hope the being returns. But the traces, only three sometimes-used fire scars, two with three stones, show those who were here come occasionally. Going from where, to where? The solitary footprint speaks of one being. For what I seek, there must be a colony. The being that was here may have been headed to its colony. Or, it may have been simply one human passing, as have travellers across time.

Today I am on foot, but I have also flown over the Barun by helicopter. This drainage is all jungle, alpine meadows, or summits. No sanctuaries show from the air—except maybe behind Saldima. Possibly the traveller heads there. That was where I was headed, sedge and grass studying on the way. Could the long slope behind the lake shown on satellite images allow entry into Tibet with the uninhabited lush Gama Valley as a food source?

To reach there from here, two high ridges must be crossed. One option is to descend to where Barun waters meet the Mangrwa, and then go up the Mangrwa to Saldima. Lendoop says that river has no trail. However, Pasang's son, Tashi, learned of the smuggler's route that goes over the two ridges. Modest profit can be had sneaking supplies across the border—from India pots and pans, spices and foodstuffs, even water buffalo skins. Once in Tibet, the trip back brings manufacturing from China, sneakers, cloth, porcelain, tape recorders, and detergent, and from Tibet brick tea, dried meat, cheese, and stud bull yaks. Such are the acknowledged trade goods.

Profit comes from savings in bribes not paid at formal borders. Second-guessing demand is complex when the commodities range from yaks to detergent. Success comes from starting with reliable, low-price suppliers and ending with quick-moving, closed-mouth buyers. Routes must handle the traffic with little breakage. The Himalaya stops high-soaring rain; its low points are easily patrolled. But one unwatched route passes through the Barun; beginning in Nepal on trails far from police checkposts in western Sankhuwasabha district. After the snow melts, the route crosses Shipton's Pass into the upper Barun, then somewhere through the jungle around me leads to the Saldima meadows, and over a 15,000-foot pass into China's uninhabited Gama Valley. From an entrepôt named Sakyathang, word goes of the smugglers' arrival. The middlemen of Kharta come by one of three passes to take the merchandise to Shegar, Tingri, Lhasa, even the heartland of China.

With today now half gone, I must find that trail. If I do not, I'll have to make my own trail across perhaps a 4,000-foot climb, probably requiring a bivouac before the top, and there's always the further unknown: the weather of Everest. I climb the first 2,000 feet in three hours bushwhacking through rhododendron. A trickle from under a rock may be the last water. Reclining to let the water absorb, accompanied by a handful of peanuts and raisins with dry grape-nuts, I watch clouds race above. Might a storm be coming? If so, it would be wise to descend to a lower altitude. Where? Everest storms close Shipton's Pass. As I think, memories are pulled off the shelf of life, breaks from prior climbs, where a partner smiled as we'd completed a hard pitch, where one bite from a snack opened flavours of distant worlds, views remembered answer the reason for climbing.

Starting climbing again, there is still no sign of a trail. Cliffs on both sides, though, speak of the wisdom of this route that appears to be a cut through the otherwise sheer wall breaking the upthrust Himalaya. Above, a treeline of scruff rhododendron looks to be easier walking. Then in from the left I come upon a line of worn rocks and slightly packed soil, a gentle contour cut by men and slipping hooves.

A trail mysterious discovered is the gathering of footfalls familiar to those who made them ... but oft an enigma to us who follow,
who wonder about those who walked before. Unlike roads in the
'civilized world' constructed by others, Himalayan trails grow from
the work of all. Porters move stones to build a stair for the next
weary leg. Herders place poles on trail sides for their animals.
Even smugglers, I now find, work on their trails. Walking on this
layering of feet that went before, the distance remaining to the
pass is covered quickly, and the total ascent turns out to be only
3,000 feet.

At the top, nowhere are prayer flags or ropes that once held
them telling that the people who use this route must be Hindus not
Buddhists. But two cairns attest to the recognition all Himalayan
peoples feel who cross from one valley to another, recognizing each
valley holds its spirits, for cairns sign a blessing for safe crossing.
An old woman once told me as we rested side-by-side at a pass:
going through a pass is like being reborn from one life world into
a next. Energies rush through like the wind, then opening to
the world you're entering. To celebrate this birthing, she said, is
the reason for the cairns; they are grateful prayers lifted one stone
upon another.

Since our son, Luke Cairn, was born, I pause at each pass, not
just to add stone to existing cairn as before, but also to start a new

15.3   Cairns at the Top of Popti La Pass

Source: Author

one. No one walks alone. If we all did not have family, we would not be here. Life comes from life travellers before, and it will be carried by those who follow. Cairns in our lives show others how to follow. I look back down the trail just ascended, at smoothed stones over which I will probably never travel again.

The valley's soil I enter is lighter, grittier, the vegetation growing from it is sparser. But it appears with the skies ahead that I do not need to worry about a storm. I am approaching the Himalaya's rain shadow. Pulling my pack's shoulder and belt straps tighter, knowing night comes in an hour, I start running down from the pass. The key in mountain running is lightness and constant flow; each stride without jolt into the next.

Downhill running Western style, however, goes foothold-to-foothold, thump-to-thump, and destroys the knees. For with each hit, the body absorbs half a ton of foot-pound pressure. Torque hammers the hard surfaces of the knee as it rebounds to take the next slam. Films of bone-end cartilage collide, and then bounce out again to absorb the next shock. The egg-white-like synovial fluid rushes in with each hit to lubricate and nourish; the knee absorber regrowing even as it is used. Forces absorbed are the most extreme any body joint takes. American boys destroy their knees playing sports. But Himalayan boys learn to cascade down rocks in lightness of running like water running. Each flying step is constant movement, each foot planted as still moving judgment for how to pivot to the next, calibrating to ricochet and never jolt the knees.

I must run almost 3,000 feet down, knowing it is wise to stop every fifteen minutes to rub my knees and refill my lungs for running at altitude burns calories faster than arteries can replenish. With fatigue is accentuated the chance to flip—and if a foot catches, the reflex must tuck, roll-in-the-air, and hit the ground on backpack. When the techniques are learned, it is addicting. All parts moving, feet only touching, coordination of the feet is accurate; pack close on my back, this time higher on the shoulders than I'd like, but it is there ready as my tortoise shell in case I flip. I run on. Suddenly I seem to feel a campfire. Feeling a smell? I cascade on, valley nearing. The camp feels like it is calling—I turn a corner and, mouths agape, stand five men, their campfire burning, with three ponies and two stud buffalo yaks.

I try to stop and, but off-balance, my legs slide out. The ponies bolt, but they have been well tied. A Nepali down the slope might be expected, but not a lone Westerner. Snarls squint towards me on the ground from these men's eyes now all around. Exhausted to the limits of my veins, I absorb that I have no exit. My pack with its cameras and instruments will exceed the value of their loads. It would take police ten days to reach this place, and they would not come unless one of these men brought them in. My remains here on the smugglers' trail—after omnivorous bears and sharp-eyed griffons—would be signature-less.

Had I thought faster, I could have—should have—bounded through this camp, letting out a piercing scream, and gone into the night leaving the apparition of a pale-skin with huge hunched back exploding through their site … vanishing. Perhaps that shock would have caused them to huddle around the campfire all night stricken with the possibility of returning ghosts, maybe even a Yeti. As I sit on my hurt tail, I know that for a forty-three-year-old fool I've made a mistake.

Nobody likes to be burst upon while settling in at night—especially people errant from the law. As shock ebbs, it seems their anger appears to rise. One of the five, who'd been splitting campfire wood, stands by holding his kukri, unsheathed.

'I'm thirsty, some tea?' I plead in a tired voice. 'Some tea?'

The oldest turns to the fire. The others stand around. The two youngest seem to seep anger. Then I realize all are just curious. I gaze through the legs standing around towards the old man taking the lid off the pot and ladling tea into a much-chipped white enamel mug, returning and offering me the mug graciously in two hands. It was I who was afraid.

'Here, I brought you raisins and American barley', I reply, unleashing pack straps and passing raisins and grape-nuts, then on tired legs wobbling to their fire. Pulling in my pack, I let my actions speak. The meal that cooks will be rice and lentils. Hoping they'll invite me, I offer peanuts and the half-full bag of M&Ms.

In the growing night, we question each other. Time comes when we must head to our blankets. Ramrod straight in my sleeping bag, my plan is to lie awake, waiting, half-expecting trouble, especially from a young man who seems to covet that which I carry. I note each of them has gone to his bed with kukri close at hand. I relax some, looking at the stars, knowing that never in my life before have I fallen asleep while lying on my back.

Waking the next morning, I am on my side having slept deeply. I've woken to the noise of trail companions preparing their loads— my fears of an exhausted earlier day now recognized as I lie in a warm sleeping bag. Smugglers is the name I gave them; they view themselves as businesspeople. Together, we travel the trails of our vocations. By the morning fire we share my last box of grape-nuts. Their direction is into the Barun to the cave in the ledge—they know of it, we talked about it last night.

My rations now total 2 pounds of peanuts, a pound of raisins, and milk powder to make three quarts. In one night my pack diminished by one box of grape-nuts, one box of raisins, half a pound of peanuts, and a half bag of M&Ms. But in return they gave knowledge of trails ahead. I know now the path to Saldima, and, if I find the trail they described, in two more nights I will make the villages north of Shyakshila where I can purchase rice and lentils. Peanuts, raisins, milk powder, and water—while walking, I work through their combinations as my next possible meals.

Seven hours later, I arrive at the hut where we had our Saldima meetings. The waterfall still explodes from above. I fill my water bottle and wait for the iodine to purify. Peanut-by-peanut, I eat. Two ravens circle. If they see food, they could burst down. In 1971, ravens attacked people eating food on Everest 2,000 feet short of

15.4   The Great Waterfall that Thunders into the Saldima Meadows

Source: Author

the summit. My eyes stay on the waterfall. No smoke rises from behind. Was that mist three years ago? Wild climbers who have gone to the top of Everest are more than four thousand—they are of my species that claims to be rational, that claims to have seen Yetis while making that climb.

ↁↂ

LAY OF TRAILS IS KNOWLEDGE IN THE MIND AS MAPS published through exchanges at resting places. Trails learned through others' tales might seem like folklore to a cartographer as they lack measurement and scale. But maps designed for the mind are to be remembered. Key crossings, these become way-points. Dramatic experiences, these are route challenges. Methods for such map making were grown long before taking to paper and presentation to scale. The smugglers had described *the idea* of the route out of the Barun, and with that I was given a from-here-to-where-I-could-purchase rice and lentils.

I knew what the idea of the trail was trying to do, and while I did not know bends and choices, those I could figure out knowing trail intent. Shared was: 'You are trying this ... and you encounter that.' A trail is understood as a live connection, not a line on the ground. How to describe a trail is also an ancient practice, different from 'go to the big maple and take a left'. In the world before writing, memory was how geography was learned, shaped by combining ideas and physical features, an extension of the purpose of travel. Trails lasted in the mind through stories.

But before going towards rice and lentils, I must climb in the opposite direction. To answer the unknown behind the waterfall I start pulling on and thrashing through the brush beside the waterfall. Saldima Meadow descends as I climb. My ascent appears to be through scrub rhododendron and thigh-high juniper. My purpose is to look down onto the lake at sunrise tomorrow, for that is when the smoke sometimes seems to rise.

When I awake, stars sparkle across the sky. Crisp juniper branches are at hand. It's tempting to lighten this still night world with a fire. But, rolling up my sleeping bag and leaving all under the overhanging boulder I slept under, I select a few emergency items and at predawn prepare to head towards the rim-crest for my view.

Stars light years above are closer than my family on the other side of the world. And the nearest settlement on this side of the world that I am sure of is three days walk through the most difficult jungle in the Himalaya. In the cold rationality of morning, I know there are no wild men in these valleys. I am alone here, in the dark of the night at 12,000 feet. Another message is also clear: for thirty years, I've been searching for a wildness that is inside me.

The rock I leave has been my one-night home. Making homes under overhanging rocks has been a human habit for millennia, epochs longer than our staying within houses. Caves are our most familiar abodes, a fact we seldom remember; branches adorning trees are what our DNA is accustomed to see as living space decorations. Then in homes built from hewn stone and carved timbers in ever increasing steps, we distance ourselves from homes of origin.

To return to a place I must again find, with its cached sleeping bag and food, I build a cairn on the rock itself. For if my return is in whiteout from that cloud that each morning rises in this valley, I need a trail to unfailingly take me to the artefacts of home I've carried here. So I use the map-building technique of seat-of-the-pants air navigation: using the land not a map held in the hand. The objective is to hit a line on the ground, like intersecting a river. Hit that wide line knowing to which side on it you need to turn, and then let that lead to the spot called home. Do not aim for the destination. And so I start writing a line out from the rock, and every 20 feet create a rock cairn. Twenty stacks later, I have a signed walkway home 100 yards long. Going back to the line's middle, I head straight up.

From the black of night a shaped world starts forming around me. As grey lightens out of the obscure, distance becomes a dimension, depth enters a world earlier seeming an endless unknown. The night's world knows no length—but as light joins it, space opens. It's about an hour still before the sun. Ice crystals chatter across my feet as frost breaks from standing grass. But the sun is coming to the rising day. And then, as I near the rim, the sun bursts over the earth's curve behind—sunsight. My shadow leaps out in front. Rainbowed prisms spring from crystals on grass tentacles that my feet tramp through.

Peering down over the rim, the lake is below. It must be almost 1,000 feet. Small trees surround its shore, a meadow too. Mist rises from its surface. Off to the right, where a gentle slope leads to Tibet's Gama Valley, smoke rises. Smoke, not mist. Against the cliff whose rim I now stand on, and along the sides of the cliff approaching this meadow, appears to be a cavity cut into the rock face, cut probably by a glacier that once moved on the meadow whose remains are the lake. Into these overhangs stones seem to have been stacked to make walls, appearing to have rooms behind them. Smoke seeps from three.

I watch through my binoculars. Will I see someone leave the shelters? Time passes, an hour then another. As I watch, from behind me the cloud has risen up the Barun, engulfed Saldima, and now rolls over the rim on which I lie. There seem to be people below. Do they discover a complementary consciousness of the yin of humanity with the yang of nature? Are these shelters their permanent home? Perhaps they may be smugglers who use this as a way station? Or, even more simply, they may be yak herders who graze that apron of grass. They also may be spiritual seekers coming closer to the great beyond. Whatever their purpose, let this community pursue its reclusive quest. By the cairns I have left, back to a world that has yet not found grace, where wildness is feared, into that confusion I shall guide my quest.

Notes

1. Jesse Oak Taylor, 'The Sky of Our Manufacture: Literature, Modernity, and the London Fog from Charles Dickens to Virginia Woolf' (Doctoral dissertation, University of Wisconsin-Madison, May 2010), p. 28.
2. Jesse Oak Taylor, 'The Sky of Our Manufacture', p. 35.
3. Jesse Oak Taylor, 'The Sky of Our Manufacture', p. 29.
4. Jesse Oak Taylor, 'The Sky of Our Manufacture', p. 31.
5. Jesse Oak Taylor, The Sky of Our Manufacture: The London Fog in British Fiction from Dickens to Woolf (Charlottesville, Virginia: University of Virginia Press, 2016).

Afterword

Previous page:
A.1 Sign and Route Map for the Yeti Trail to Lead Visitors into the Barun Valley

Source: Author

MAY 2010. I RETURNED TO SALDIMA MEADOWS. Accompanying me were my two sons both now men. Jesse Oak was returning to jungles where in life-opening ways he first entered the wild. Luke Cairn came to the mountains he had been travelling in, by then for years, doing his own research. We had returned because my Yeti quest continues.

Two Yetis exist. Each has a different identity. The maker of the footprints is a bear; that identity is certain. Beyond the footprint maker, though, is a second Yeti, one asking existential questions about Homo sapiens' relationship with the wild, and those questions each person needs to answer individually. To help with that, in the Yeti a symbol is given with which to discover one's own footprints—this is far from an abominable search.

The footprint-making Yeti's spoor I have tracked across mountains, seen its nests in the high trees, and watched it feed. Tranquilizing it, I replicated the footprints in plaster to match the earlier mysteries found by others in the snow. As hunger calls (or urge to reproduce) from one side of a mountain this bear, *Ursus thibetanus*, goes over the mountain. Its prints then emboss in glaciers. That explanation, simple as it is, fits all the facts.

Yet the enigma continues, for the Yeti has a second identity that is more than a bear. This is a mascot that walks the world only loosely tied to the Himalaya. What is extraordinary about this reality is that it lives not in the snows but in human desire. This is not a physical animal. People believe in this Yeti as an embodiment of the human connection to the wild. As an icon represents faith and as an idol symbolizes an idea, so is this second Yeti both an icon and an idol.

What is extraordinary about the Yeti that exists as a bear is that from that enigma whose trail cannot be followed, where the footprints themselves melt the next day, from this resulted real national parks. Moreover, these parks brought a new way to manage the wild by people caring for the wild. The parks that were started by this Yeti are: the Makalu–Barun National Park in Nepal and Qomolangma (Mount Everest) National Nature Preserve in China. The model started there was adapted across the Himalaya in Nepal, China, India, Bhutan, Myanmar, and other places. What resulted across the Himalayan span was protection for the bear and other species.

But beyond, these homes have become havens for the human desire to connect with a primordial wild, the second Yeti. These collected strings of national parks have become an expansive diverse place where people can live in some balance with the wild.

When established in the 1980s, our two partnership-based parks were the first people-based nature preserves in Asia, opening the idea of community-engaged partners that had before been only talked about such in the world park conference in 1982 in Bali. The wild's preservation by engaging people in its management became real. This gives hope. For against that hope the wild as we once planet-wide knew it disappears. The mystery still to be solved is whether the species that always hungers for more will turn now planet-wide to collectively manage what it has taken. Will we grow a shared commons for generations yet to come?[1]

In 2010, I made that first trip back to Saldima, and then four months later I made a second trip. They both were prompted because I had lost my father a few months before, and so I went into nature asking existential questions. Walking in these mountains, in which Dad had been born almost a century earlier, I first took my sons along, then went alone, probing through protected valleys in his memory. These pilgrimages opened connection. Families are animations of the individual, extensions and gatherings from where we have come and giving membership to where our kind continues.

Travelled on my two pilgrimages were trails intimate with towering trees, with feet splashing through pools at the base of waterfalls bringing melted snows from Himalayan summits. In those life journeys my footfalls treaded paths few people know, edged around the shoulders of earth's first, fourth, and fifth highest mountains—where above the planet penetrates space, the one place earth goes further towards beyond our knowing.

Regrowing natural luxury gives hope for the world—for creating the national parks has established baselines of successful action to regrow nature. Against that baseline, comparing from those, we can show animal and plant numbers now more than when the action began. We need not head towards a planetary loss. We have evidence that people's behaviours changed in this region, evidence that the species that typically damages can do the right thing.

Humans can adopt modalities of living to make the wild wilder. To reinforce this lesson, I turn again to Jack Turner's *The Abstract Wild*: 'A place is wild when its order is created according to its own principles of organization—when it is self-willed land'.[2] This type of wild is not one that is managed (the US Government's Wilderness Act, for example) where control is by policies, licenses, and not admitting mistakes. Nor is this the wild promoted when the focus is on biodiversity priorities (many conservation groups or biologists, for example). This type of wild is more, a wild that rises from priorities inside people. That wild is a process not an end.

The new mystery coming from the Yeti is a 'whether'. Are we allowing human footprints to trend towards where 'order is created according to its own principles'? This means letting bears come into our cities (of course, taking them out when they harm us). It also means letting people organize their lives (rather than organizing it for them). The new mystery will be created by working through this question: Do we seek balance or do we seek control in our lives?

It is impossible to control the experience of living. The forces of life are too large, our wisdom too small. People everywhere are increasingly worried by changes underway in the united socio-econo-info-biosphere that our planet has become. Some in reaction gravitate to doctrinal fundamentals, hoping in ardent belief questions shall go away. But others are absorbing; we are part of a large complex always adapting system. The wild is not only a feature of the past ... it can become a fact of Homo sapiens' future (people who are wise).

Learning to live in the *idea of the wild* is, like this, a path. The opening way as we go towards the unknown is to embrace the wild and walk among it. With adjusted behaviours, we grow wildness again. This is our foundational calling as we enter the age of human-making of climate, disease, civil strife, and economic surprises. The opportunity is about growing a new earth. More flowers were not the meadow's earlier condition, but in the world of our making it is evolution's potential.

On my second pilgrimage through Everest, Makalu, and Lhotse's valleys, alone, living with the contents in my pack, I saw no other human. My life was filled with forces that played music from nature as does wind when careening through alpine gorges, for valleys are colossal flutes that can grow sonorous—and from great mountains thunder avalanches that resonate in valley chambers below. Majestic—what other word describes such an event—so close to magic for that is how it feels. Against such a backdrop rose trees 200 feet high and 6 feet in diameter. While behind those, Everest and her sisters towered 15,000 times higher. Answers that came were more than Latin names and ecological niches. Underway here are people making the planet wilder *now*.

This rocky chunk flying through space, a hardness totally devoid of life two billion years ago, somehow grew great life, and one species among that life became planet reshaping. Religions give differing answers to this. But in a new reshaping, on our flying chunk's highest place and on one of its more fragile ecosystems it was possible for life to become wilder than when, a third of a century before, the abominable search began. The wild was not pocketed off from people as in the old mode. Living with the wild was brought into domesticated lives. Despite India and China growing with their billions—and counter-intuitively because of a wild human now shown not to exist—a life repository grows wilder because of increasing numbers of people coming into more active engagement.

What is important to realize is that the wildness has always been with us, relentlessly. So, let us invite it in, learn to live with it, and enter into a dialogue with the ground of all being that which gives life. Icon and idol—that is the discovery the Yeti led me to.

ॐ

BUT HOW TO EXPLAIN THE FOOTPRINTS THAT LAUNCHED THE SEARCH? A half century of continued findings confirms the bear explanation.[3] After my first book explaining the tree bear being the Yeti, the explanation was repeated around the world. A range of discussants concur that a bear makes these footprints. The mountaineer Reinhold Messner came to this conclusion. So did Disney World, of all places, and the consensus grows.[4] The bear explanation, as noted before, is not original to me; it was first advanced by Smythe in 1937 then reaffirmed by Charles Evans in 1955.[5] While agreement coalesces on the bear, until my explanation no point-by-point elucidation explained the specific Shipton print that launched Yeti interest worldwide.

Ursus thibetanus, the Asiatic black bear, made that elongated 1951 footprint. When its front paw came down, making those dramatic toes in the snow, it did not press so firmly into the crust to show the bear's nails on the front paw. Then the hind paw fell onto the back half of the print, elongating to the twelve-inch length and from the hind paw nail marks are evident—and with this feature the story gets more interesting.

The four-inch shorter 1972 prints of Cronin and McNeely, Tombazi three decades earlier, and our discovery that day on the ridge were all made where the hind paw came down with less of an overprint.[6] We have then a 'smaller Yeti'. A fact overlooked by others that proves the Yeti to be a product of misidentification is that no credible Yeti prints have ever been photographed of a Yeti going downhill. That going uphill is needed to make Yeti footprints is evidence that the mountain makes the Yeti, not the animal. Steep hills make big Yetis, and almost flat slopes make small Yetis.

The hominoid-like thumbs evident on some footprints are created by young bears who have dropped inner digits pressed down on the paw (yearlings and twolings have agile tendons and joints) because they spend much of their time in the trees (becoming therefore rukh balu, tree bears) seeking food. And these dropped inner digits can look remarkably like a thumb when imprinted in snow.

But let us look specifically at that iconic 1951 Shipton and Ward image, showing how *Ursus thibetanus* is the maker of this mystery that started it all:

> Nail marks of the hind paw are revealed *in the center* of the overlay print. With most observers' attentions on toe pads at the top of the print the determinative feature of the print's maker are

two nail marks *in its center* (one on the right side and the other on the left).

The mind-captivating human-like toes on the top that do not reveal nails are created because the front of the bear is less heavy, causing weight not to push the front paw as deeply into the snow as the rear. (Cronin and McNeely's snow was soft, and so the nail marks did not show.)

The three tipped-to-one-side toes at the top of the print align almost identically with the digits of an *Ursus thibetanus* front paw while the 'thumb' on the left fits with a tree bear's splayed inner digit. (Bears typically walk bow-legged.)

And, on that Shipton print, when the rear paw came down partway down the print, the nail marks are in the center of the print, an unexpected place for nails if the print was made by one foot but not if it is a second foot. That the rear paws penetrated more deeply is because the rear of the bear is heavier than the front.

Because Shipton mentions 'for there were several', also suggests bear. For it is likely these prints are from one or two cubs accompanying the mother—evidence shown earlier in the photograph of Michael Ward standing beside their mysterious Yeti trail.

The explanation of known animal for mysterious Yeti may also include the snow leopard's eerie yowl which from time to time reverberates off high Himalayan walls, an animal very rare in Sherpa country in the middle of the last century where the Yeti legend grew. But with recent conservation, its numbers increase so the snow leopard is now known to have a range of vocalizations.

I had come to this Yeti/bear conclusion after sixty years of research (1956–2016). The explanation fits all the facts. Then something remarkable happened as this book was going to the press. I had contacted the Royal Geographical Society for permission for Oxford University Press to publish the iconic Shipton print I had carefully studied for so many years. The society came back asking, 'Which Shipton Yeti photograph do you want?' They sent two images. In my sixty years I had never seen their second one. See A.2.

A.2   The New, Never-Before-Published 1950 Shipton Yeti Photograph

Source: Royal Geographical Society

Photographed is the same print as the familiar one shown in Chapter Four (see 4.2). But this new print, taken a bit further away, has three new details. First, are two nail marks on at the top of the lower partial footprint; nail marks exactly of the expected dimension between the second and third digits for *Ursus arctos thibetanus*. Second, between the familiar print and now the partial print seen below are three scratch marks; I suggest these marks were made by the bear's front foot just before it put that foot down. They are possibly rear foot prints, but what is certain is that they are bear nail marks. A third point is of interest; the icy crust was indeed very thin, explaining why the bear's feet did not sink in, because at the top of this new print the rock beneath is evident.

So again I argue, the Yeti is a bear. This new print provides added proof. Nonetheless, across now three decades since first making the identification, in letters I receive, questions after lectures, and call-ins during radio shows, the Yeti's bear identity is not what people focus on. The Yeti lives in the larger ideology. It is a mascot suit, and inside this suit is a human hunger. A second Yeti exists: the hope that there *might be* a connection today to eons gone by.

Indeed, that desire is as accurate an answer as that for the first Yeti. Wildness is disappearing. It does not matter that the Yeti is a bear that clambers out of jungles and crosses high passes. To people who hunger for the wild, what matters is to have alive a mystery from the frontier of the planet—reminding us that, in the Anthropocene, wildness is still possible. What helps in this new age of human making is a hope that guides our way as we apprehend the frighteningly changed wild that is coming.

⟲

AS A HAVEN FOR THE WILD, PEOPLE-BASED ACTION IS THE HOPE. To preserve the potential for the wild in people's lives, evidence accumulates that conservation is more effective, less costly, and more sustainable when done in partnership with people. In Asia, the approach was initiated by Makalu–Barun National Park, then the Qomolangma (Mount Everest) National Nature Preserve.

Despite a civil war in Nepal that burnt park buildings and drove out park wardens, today forests and wildlife are more abundant than when we 'discovered' the pristine Barun. And in China's Tibet Autonomous Region, a land which public opinion holds erroneously to be undergoing devastation, wildness also grows. In both areas, what has been successful is protection that advances through actions by communities in partnership with the government and science.

In Nepal, communities did not take advantage of the war to invade the Barun. They did not just control themselves (which they did); they also prevented exploitation by others. The 20 acres of jungle where the Barun and Arun Rivers meet is proof. Decades earlier, Tirtha identified these acres as a key subtropical habitat, where warm, moist air is pulled up the Arun Valley by the low-pressure zone off Everest and her sisters to nurture this niche in the middle of the Himalaya. When our expeditions came, the people were turning this jungle into fields (it was here Lendoop's daughter slipped into the river). But the people stopped clearing that jungle. Today those acres remade by human actions are fitting the authentic original Sanskrit meaning of jungle: fields returned to the wild.

Shyakshila gives an example of how a Nepal village many would view as so poor that to advance it needs 'assistance' was able to do 'the right thing'. Getting them to advance did not come from buying their actions or from a leader telling them what to do. While employing a people may get those individuals to comply with legal ordinances, and power may turn people to protection as long as power is exerted, what Shyakshila showed is how the people learned—and from that adopted a partnership with nature. Learning is what changes behaviours sustainably and at scale.

The larger Nepal experience confirms the effectiveness of this people-based approach. A study was done with the Annapurna Conservation Area whose policies were shaped the year after the Saldima Meadow conference. The study found that twenty years after Annapurna's creation while only 15 per cent of the population perceived economic benefit from the park's main revenue source of tourism and eighty-five per cent reported hardship due to loss of crops or livestock from wildlife, support for the preserve was high. They recognized receiving other benefits such as infrastructure and services.[7] In addition to these other benefits, they had learned and expanded their understanding. The Annapurna people like those of Shyakshila had pride (as do so many people around the world when they can live with nature).

When we learn to live in something we had never noticed before and with all its interlinked parts—the socio-econo-info-biosphere—a dimension of wholeness grows in our lives. This is more than conservation (using less) and more than protection (not using). Conservation or protection often begins in the premise of taking land off-limits.[8] Doing so robs people of connection. People do not voluntarily join in working together when seeing things taken from them. What works is to adopt the mindset of *being part of*. Effectiveness comes in mutuality.

Nepal's community forestry experience also proves this. Historically, Nepal's forests belonged to government or large landowners. As Nepal's population trebled, villagers stole timber as they needed wood. As the villagers were winning, it was not just the forests that were losing but also the national life. Then, in 1976, forests began to be transferred to Community Forest User Groups. Over two decades 17,700 groups formed and 1,650,000 hectares of managed forests came into being. Now one out of every three Nepali citizens is a member of a user group. From shared forests the people get fuel wood, fodder, grass, non-timber forest products, and poles for home construction.[9] Other benefits come too, such as breaking down caste and wealth barriers, diversifying governance, and expanding income and credit opportunities.

Forests too benefitted. Nepal's once denuding hills, which I remember prognosticating in 1970, would be bare red earth, today grow green across the country. My prediction was wrong because I saw people as the problem. A representative study of eleven user groups found that forest cover and biodiversity has improved in all eleven research sites.[10] Particularly significant is how through the ten-year civil war, the forests continued to be maintained even when the rebels were using the forests as military redoubts.[11] To enable the growth of trees, communities come together.

Parallel proof is in China's Tibet Autonomous Region. The Qomolangma (Mount Everest) National Nature Preserve spawned thirteen other protected areas, conserving 44 per cent of the land area of Tibet.[12] (More recently, that growth has expanded further to 54 per cent of Tibet and eighteen preserves.) A 1,200-acre preserve was even established in the city of Lhasa, bringing the wild into the heart of the city, right up to the valuable urban land of the Potala Palace.[13] Across the Tibet Autonomous Region, each preserve is managed by community-engaged systems that follow with adaptations the Qomolangma (Mount Everest) National Nature Preserve model. Every species of wild animal, when assessed at the Tibet scale, has population numbers rising. There is a regeneration also of forests on the lands outside the half of the land formally protected.[14]

ॐ

SO AT THE END OF THE SEARCH FOR A WILD MAN IN THE SNOWS, a new wild grows. As a result, living with people are snow leopard, wild ass, musk deer, four types of wild sheep, rare pheasants, dozens of species of rhododendron, the majority of species of earth's poppies and primulas. Many interesting aspects of life are included in this such as the giant cypress, one of earth's largest and least known trees found only in the Tsangpo River gorge between the high-water flood mark and the riverbank.

On exiting from my solitary pilgrimage through these valleys, I met old friends Lendoop and Myang after a gap of a quarter of a century. Walking into Shyakshila, these men now old literally picked me off the ground with their colleagues and threw me back and forth. As we sat again on floors of their homes eating boiled eggs and drinking tea, Lendoop asked for help to make a walking trail into the Barun jungle. I was shocked. Of all places, the Barun is sacred—it must be kept unpeopled. While I can advance people-based management as national policy, certain areas should be apart I thought. And, of course, the cathedral of Nepal's wildness was holy: the Barun.

Sitting on the floor where years before I had slept, neither the home's floor nor wealth had visibly changed, and still it was these people who, to external material perspective had not advanced, had brought the wild back. Lendoop explained that because of the new park and the strict protection given to the Barun, now tourists walked around the Barun Valley when they went to see the mountains. As a result, the people of Shyakshila were losing employment, and, they noted, the tourists missed seeing the special jungle they had conserved for then a quarter of a century and through a civil war.

Embodying holy meaning, making the sacred sacred, I realized, comes from people valuing that above themselves. An idol, a crucifix, or a mosque becomes sacred when people lift it up to be so. The opportunity Lendoop was proposing was for the world to join in making the Barun sacred. To place their vision into my words, that would happen when people from outside could walk into the sacred space. People would enter the holy of Divine organization when they walked into the Barun—making the proposed trail would enable that—as in cathedrals an aisle allowing people to approach the alter makes profound the act of devotion. Maybe my science-minded colleagues and I in emphasizing action that denied access three decades had erred.

What creates authentic preservation? I could raise money and hire people to protect the land. I could even raise money to preserve Shyakshila as a 'historic village' that displayed traditional life. The Barun and Shyakshila would then be 'preserved'. But if every human holds equal right to partake of the world, Shyakshila should not be held in the past. Shyakshila should not just be invited into the process, but could help lead others.

So I reached into my bank account and gave several thousand dollars to purchase axes and shovels. Friends added more thousands to cleave *The Yeti Trail* through the jungle that today transects the Barun Valley (see A.1 and A.3). What is being created is pristine jungle connected to the world. The natural cathedral has that aisle now up its centre, and the people of Shyakshila are this basilica's sextons.

A.3 The People of Shyakshila Village Making the Yeti Trail through the Barun Jungles

Source: Author

We have the opportunity to grow the wild back. In this wild-ness is the preservation of the world. Because, when (and if) the Yeti is proven to not exist, what will happen? Humans will have lost a travelling connection for the journey of human experience. Humans will live, then, in separation from life. And, in this nar-rowly made creation, we shall have lost our genealogy.

This new wild of human making is more dangerous than the natural wild we once lived with. Homo sapiens today prowl in a world more perilous than that of tigers, cobras, and bears. That approach that separated us from the wild, saw nature as fearful, and changed the planet's age from Holocene to Anthropocene. Unless we now change ourselves, we risk recoding our species in aspects more determinative than DNA from *sapiens* (wise) to *Homo arroganticus.*

Postscript

This narrative of my Yeti journey is factual with place, time, and discovery—except in one instance. The frozen lake behind the waterfall above Saldima is as described in Chapter 15. But it is not at that place where the community lives that was discovered with their homes adjacent to a glacial meadow. Nor does this com-munity live in the Barun or Gama Valleys. I am not revealing the location, but there is such a community in the Himalaya, indeed I know of several, and their story is woven in here to recognize people that are now intentionally moving from privilege into living with the wild.

To find them we do not need to search the Himalaya. The Hindi word for them is sadhu, seekers of moksha—liberation. Such people circle the world. On the Chinese side, these people are called xian. We live in a planet-connecting community of individuals learning to live with rather than take from. The communities they create exist everywhere—each of us has the opportunity in our lives with our neighbours.

In my distinctive life journey, I have been fortunate. All children grow up with spirit worlds—goblins, bogeymen, and spectres that give persona to incomprehensible natural forces. These characterizations explain the forces we naturally sense, feel very alive in our imaginations. One explanation for the Yeti is just that, persona to incomprehensive natural forces—where, for some people, wild hominoid understanding will always be.

I have been blessed, though, to grow up with a different Yeti. Convinced of a wild human life, I followed it through splendid valleys. I grew from this engagement so that with many colleagues from many countries we worked, and from that shared work brought forward a way. I am not a sadhu, but I have been privileged many times to touch moksha. Being with the Divine is joined through prayer as well as the work of service. Such a possibility is there for all, those lucky to know the wild, as well as those who imagine the

A.4   The Author in an Ice Cave at 19,000 Feet in Gosainkund Himalaya

Source: Lorenz Perincoli

wild. We thrive within these opportunities if we let them open to probe the deepest meanings of what it means to be alive.

Notes

1. I have written two academic accounts of the growth of this people-based conservation approach, especially its dramatic extension across the Tibet Autonomous Region of China. See Daniel C. Taylor and Carl E. Taylor, *Just and Lasting Change: When Communities Own their Futures, 2nd Edition* (Baltimore MD: Johns Hopkins University Press, 2016), Chapter 20, pp. 274–89.

   Also, Daniel C. Taylor, Carl E. Taylor, and Jesse O. Taylor, *Empowerment on an Unstable Planet: From Seeds of Human Energy to a Scale of Global Change* (New York: Oxford University Press, 2012), Chapter 8, pp. 157–86.

2. Jack Turner, *The Abstract Wild* (Tucson, Arizona: The University of Arizona Press, 1996), p. 112.

3. I first published these findings in 1995: *Something Hidden Behind the Ranges* (San Francisco: Mercury House Press).

4. Answers continually come from surprising corners. In the spring of 2014, Professor Bryan Sykes at the University of Oxford studied 32 samples of alleged Yeti and Bigfoot hair using DNA analysis. His conclusion: most were dog, horse, deer, and the two most mysterious were bear and with these his conclusion was the improbable, totally out-of-range polar bear and the more likely *Ursus arctos* (brown bear). But he then went on to draw a conclusion that this was not a regular brown bear but 'an anomaly'. Again, the proclivity to connect to 'an anomaly' may be driven by emotional gravitation to Yeti rather than the simpler bear identification. Bryan C. Sykes, Rhettman A. Mullis, Christophe Hagenmuller, Terry W. Melton, Michael Satori, 'Genetic analysis of hair samples attributed to yeti, bigfoot, and other anomalous primates', in *Proceedings of the Royal Society*, B281: 20140161. Recovered 26 February 2017: http://rspb.royalsocietypublishing.org/content/royprsb/281/1789/20140161.full.pdf.

   A refutation of this claim on scientific grounds is presented in: Eliecer E. Gutierrez and Ronald H. Pine, 'No Need to Replace an

"Anomalous" primate (Primates) with an "anomalous" bear (Carnivora Ursidae)', published in *Zoo Keys* 487: 141–54; 15 March 2015.

5. A parallel review of most of the Yeti claims to that in this book is found in Daniel Loxton and Donald R. Prothero, *Abominable Science* (New York: Columbia University Press, 2013). They proceed through the more-than-a-century-long history of claims, and conclude in a similar way with the bear being the primary footprint maker.

6. I subsequently have found midsize prints between the Cronin and McNeely, and Shipton and Ward, but leave them out of this narrative because they do not fit the story flow. I have also since found overprinted 'Yeti' prints in mud, but they are so clearly of a bear that it is easy to see why across the decades no Yetis have been 'discovered' outside the snows.

7. Siddhartha Bajracharya, Peter Furley, and Adrian Newton, 'Impacts of Community-based Conservation on Local Communities in the Annapurna Conservation Area, Nepal', 2005, *Environmental Conservation* 32(2): 239–47.

8. Mac Chapin, 'Can We Protect Natural Habitats without Abusing the People who Live in Them?' *World Watch Magazine*, November–December, 2004.

9. Michael Rechlin, Daniel Taylor, Jim Lichatowich, Parakh Hoon, Beberly de Leon, and Jesse Taylor, *Community-based Conservation: Is it more effective, efficient, and sustainable* (Franklin, WV: Future Generations Graduate School, 2008), p. 57.

10. N.P. Yadav and O.P. Dev, 2003, 'Forest Management and Utilization under Community Forestry', *Journal of Forest and Livelihood* 3(1): 37–50.

11. Michael Reclin, Bill Burch, Bhishma Subedi, Surya Binayee, and Indu Sapkota, 'Lal Salam and Hario Ban: The effects of the Maoist insurgency on community forestry in Nepal', 2007, *Forests Trees and Livelihoods* 17(3): 245–53.

12. Daniel C. Taylor and Carl E. Taylor, *Just and Lasting Change: When Communities Own Their Futures, 2nd Edition* (Baltimore, MD: Johns Hopkins University Press, 2016), pp. 285–89.

13. Taylor and Taylor, *Just and Lasting Change*, p. 288.

14. Robert L. Fleming, Liu Wulin, and Dorje Tsering, *Across the Tibetan Plateau: Ecosystems, Wildlife, and Conservation* (New York: W.W. Norton, 2007), p. 6.

Glossary

Bhui balu	ground bear
chang	barley beer
Deodar	Sanskrit for 'wood of God'
Ghoral	the Himalayan chamois
Kali Khola	stream of the Goddess Kali, the goddess of destruction
needeene	local name for human-like jungle spirits
Payreeni Khola	landslide stream
po gamo	local name for human-like jungle spirits
rakshi	distilled liquor made of rice or millet
rukh balu	tree bear
Serow	a donkey-sized, red-brown mountain goat with white legs
shockpa	local names for human-like jungle spirits
thangka paintings	Traditional Tibetan iconography
Thar	large golden-fleeced Himalayan mountain goat

About the Author

Daniel C. Taylor has been engaged in social change and conservation worldwide for four decades with a focus on building international cooperation to achieve ambitious projects. He founded the seven Future Generations organizations worldwide, including the accredited Future Generations University (www.future.org, www. future.edu). He also co-founded and led The Mountain Institute (www.mountain.org) and initiated its worldwide programmes. In 1985, after providing the scientific explanation for the Yeti, he led a scientific team creating Nepal's Makalu–Barun National Park, followed by, in close partnership with the Tibet Autonomous Region, China's Qomolangma (Everest) National Nature Preserve, then Lhasa Wetlands National Park, and Four Great Rivers Nature Preserve.

Daniel is one of the synthesizers of the SEED-SCALE method, an understanding of social change initiated by a UNICEF task force he co-chaired from 1992 to 1995. Since 1995 he continued to lead global field trials of SEED-SCALE and is senior author of *Just and Lasting Change: How Communities Own Their Futures*, 2nd edition (Johns Hopkins University Press, 2016; 1st edition, 2002) and *Empowerment on an Unstable Planet: From Seeds of Human*

Energy to a Scale of Global Change (Oxford University Press, 2012). He is also the author of four other books on various topics.

Daniel, whose childhood was in the Himalaya, received his masters and doctoral degrees from Harvard University. From 1969 to 1972 he worked for the U.S. State Department. Among many honours, Daniel was knighted by the King of Nepal *Gorkha Dakshin Bau* III; was made the first Honorary Professor of Quantitative Ecology of the Chinese Academy of Sciences; and was decorated with the Order of the Golden Ark by HRH Prince Bernhard of The Netherlands.